39

MICROBIOLOGICAL RISK
ASSESSMENT SERIES

Control measures for Shiga toxin-producing *Escherichia coli* (STEC) associated with meat and dairy products

Meeting report

Food and Agriculture Organization of the United Nations
World Health Organization

Rome, 2022

Required citation:

FAO and WHO. 2022. *Control measures for Shiga toxin-producing* Escherichia coli (STEC) *associated with meat and dairy products – Meeting report.* Microbiological Risk Assessment Series No. 39. Rome. https://doi.org/10.4060/cc2402en

ISSN 1726-5274 [Print]
ISSN 1728-0605 [Online]

ISBN [FAO] 978-92-5-136984-5
ISBN [WHO] 978-92-4-005857-6 (electronic version)
SBN [WHO] 978-92-4-005858-3 (print version)

Cover picture ©Dennis Kunkel Microscopy, Inc.

Layout: Tomaso Lezzi

Contents

ANNEXES

Acknowledgements

The Food and Agriculture Organization of the United Nations (FAO) and the World Health Organization (WHO) would like to express their appreciation to all those who contributed to the preparation of this report through the provision of their time and expertise, data and other relevant information before, during and after the meeting. Special appreciation is extended to all the members of the expert committee for their dedication to this project, in particular, to Todd Callaway for his leadership in chairing of the meeting and his support in preparing the final report. All contributors are listed in the following pages.

The preparatory work and expert meeting convened to prepare this report was coordinated by the Secretariat of the Joint FAO/WHO Expert Meetings on Microbiological Risk Assessment (JEMRA).

CONTROL MEASURES FOR SHIGA TOXIN-PRODUCING *ESCHERICHIA COLI* (STEC)
ASSOCIATED WITH MEAT AND DAIRY PRODUCTS

Contributors

EXPERTS

Frederic Auvray, National Veterinary School of Toulouse, France

Li Bai, China National Center for Food Safety Risk Assessment (CFSA), China

Todd Callaway, University of Georgia, the United States of America

Natalia Cernicchiaro, Kansas State University, the United States of America

Isabel Chinen, National Infectious Disease Institute, Argentina

Paul Cook, Food Standards Agency, the United Kingdom of Great Britain and Northern Ireland

Roger Cook, Ministry of Primary Industries, New Zealand

Patricia Desmarchelier, Food Safety Specialist / Consultant, Australia

Geraldine Duffy, Food Research Centre, Teagasc, Ireland

Peter Feng, US Food and Drug Administration (Retired), the United States of America

Jorg Hummerjohann, Bacteriological Food Safety, Agroscope, Switzerland

Ian Jenson, Meat & Livestock Australia, Australia

Musafiri Karama, University of Pretoria, South Africa

Tim McAllister, Agriculture and Agri-Food Canada, Canada

Sara Monteiro Pires, The National Food Institute, Denmark

Camilla Sekse, Norwegian Veterinary Institute, Norway

Roberto Vidal, Universidad de Chile, Chile

RESOURCE PERSONS

Sarah Cahill, Joint FAO/WHO Food Standards Programme, Italy

Verna Carolissen-Mackay, Joint FAO/WHO Food Standards Programme, Italy

Charmaine Chng, World Organisation for Animal Health (WOAH), France

Jose Emilio Esteban, Codex Committee on Food Hygiene, the United States of America

Goro Maruno, Joint FAO/WHO Food Standards Programme, Italy

Jenny Scott, Codex Committee on Food Hygiene, the United States of America

Hajime Toyofuku, Yamaguchi University, Japan

Constanza Vergara, Codex Committee on Food Hygiene, Chile

Lingping Zhang, Joint FAO/WHO Food Standards Programme Food, Italy

SECRETARIAT

Haruka Igarashi, World Health Organization, Switzerland

Christine Kopko, Food and Agriculture Organization of the United Nations, Italy

Jeffrey LeJeune, Food and Agriculture Organization of the United Nations, Italy

Satoko Murakami, World Health Organization, Switzerland

Kang Zhou, Food and Agriculture Organization of the United Nations, Italy

Declaration of interests

All participants completed a Declaration of Interests form in advance of the meeting. The Interests declared were not considered by FAO and WHO to present any conflict in light of the objectives of the meeting.

All the declarations, together with any updates, were made known and available to all the participants at the beginning of the meeting. All the experts participated in their individual capacities and not as representatives of their countries, governments or organizations.

Abbreviations and acronyms

CAC	Codex Alimentarius Commission
CCFH	Codex Committee on Food Hygiene
CFU	colony forming units
CRISPR	clustered regularly interspaced short palindromic repeats
DALY	disability-adjusted life year
DDGS	dried distiller's grains with solubles
DNA	deoxyribonucleic acid
eae	*Escherichia coli* attaching and effacing gene
EFSA	European Food Safety Authority
ELISA	enzyme-linked immunosorbent assay
EPEC	enteropathogenic *Escherichia coli*
FAO	Food and Agriculture Organization of the United Nations
GAP	good agricultural practice
GHP	good hygiene practice
GMP	good manufacturing practice
HACCP	hazard analysis and critical control point [system]
HUS	haemolytic uremic syndrome
JEMRA	Joint FAO/WHO Expert Meetings on Microbiological Risk Assessment
NTS	non-type specific
PCR	polymerase chain reaction
PFU	plaque forming units
SCFA	short-chain fatty acids
STEC	Shiga toxin-producing *Escherichia coli*
stx	Shiga toxin gene
Stx	Shiga toxin
TVC	total viable bacterial counts
US FDA	United States of America Food and Drug Administration
USDA	United States of America Department of Agriculture
VTEC	vero toxin-producing *Escherichia coli*
WDGS	wet distiller's grains with solubles
WGS	whole genome sequencing
WHO	World Health Organization
WOAH	World Organisation for Animal Health

Executive summary

Shiga toxin-producing *Escherichia coli* (STEC) are estimated to cause more than 1.2 million illnesses and 128 deaths globally each year. The previous work of FAO and WHO identified beef and other types of meats, dairy products and produce as significant risk factors for STEC infection. As such, at its 42nd Session, the Codex Alimentarius Commission (CAC) endorsed the Codex Committee on Food Hygiene's (CCFH) recommendation for the development of guidelines for the control of STEC in beef, raw milk and cheese produced from raw milk, leafy greens and sprouts. To facilitate this work, the CCFH requested that FAO and WHO Joint Expert Meeting on Microbiological Risk Assessment (JEMRA) provide scientific advice on the effectiveness and utility of control measures against STEC during primary production, processing and post-processing of raw meat, raw milk and raw milk cheeses.

During the meeting, the expert committee reviewed interventions for the control of STEC in cattle, raw beef and raw milk and raw milk cheese manufactured from cows' milk, and also evaluated available evidence for other small ruminants (goat, sheep), swine and other animals (reindeer, yak, camelids, bison, buffalo and swine). The expert committee was tasked with scoring the degree of support for the effectiveness of interventions for the specific control of STEC as high, medium or low based on the evidence available within the scientific literature.

In meat production and processing systems, many approaches to support control of STEC are based on good agricultural practices (GAP) and/or good hygiene practice (GHP) that aim to generally reduce the spread of pathogens and are not specifically focused on STEC. On-farm, these include managing the hygienic conditions of housing, bedding and drinking water hygiene, appropriate animal density and biosecurity measures, effective sanitation of facilities and proper disposal of manure.

On-farm, several dietary and herd management strategies with varying levels of impact on STEC populations in beef and dairy animals have been explored. Evidence to support cattle demography (Section 2.1.3), animal density (Section 2.2.2), biosecurity (Section 2.2.1), and environmental hygiene (Section 2.2.3) were rated as having a medium or medium to high degree of support with regards to their ability to impact STEC. Interventions including feeding of forage versus concentrate rations, specific grain types (Section 2.3.3), and the inclusion of citrus products and essential oils in feed (Section 4.2.5) were supported at low to medium or medium degree of support, yet probiotics may be useful with administered to cattle, goats and sheep through feed (Sections 2.3.4.1 and 6.1.1). Some vaccines have been shown to reduce faecal excretion of STEC O157:H7 (Section 2.4.1), but their efficacy is variable depending on the vaccine and the number of doses administered.

Long distance transport and the stress of interim unloading/loading have been shown to increase faecal excretion of STEC that can lead to cross-contamination between animals (Section 2.6). Transport distances should be minimized in accordance with best practices for animal welfare, and the evidence related specifically to the control of STEC was supported at a low degree. A summary of primary production control measures for STEC in cattle and their degree of support rating (high, medium, low), based on scientific evidence, is available in Annex 1.

Avoiding contamination of the carcass through contact with hides, gut contents or faeces during slaughter is an accepted management practice during meat processing, but evidence supporting the effectiveness and reliability of these measures for the control of STEC was limited. Processing measures where evidence supported a high or medium to high rating for efficacy in STEC reduction included steam vacuuming of visible faecal contamination on carcasses (Section 3.3.4.3), and the use of a hot potable water carcass wash, steam pasteurization followed by 24 h air chilling and combinations of these (Section 3.4). The use of knife trimming to remove carcass tissue contaminated with faecal material is common and is supported by a medium confidence level in the evidence (Section 3.3.4.2). Despite the commercial use of pre-chill carcass decontamination treatments using organic acids and other chemical agents, the confidence in the evidence was low in cattle and other small ruminants due to high variability in results (Section 3.4.3). A summary of processing control measures for STEC in beef and their degree of support (high, medium, low), based on scientific evidence, is available in Annex 2.

The efficacy of available control measures for reducing or eliminating STEC on primal cuts, trim, cheek meats, and ground beef was widely varied. Yet, the use chemical antimicrobial dips (Section 4.2) for primals and trims were supported at a low to medium level of confidence, and high-pressure processing (HPP) (Section 4.1.6), gamma irradiation and electron beam sterilization (eBeam) (Section 4.1.7) produced significant reductions of STEC in ground beef and in retail packs. A summary of post-processing control measures, and combinations of these, for STEC in beef and their degree of support (high, medium, low), based on scientific evidence, is available in Annex 3.

Pork products and meat from wild game have occasionally been confirmed as vehicles of STEC transmission, but there are no interventions or practices during the processing of these animals that are specific for STEC. Meat from these species could be treated post-harvest in a similar fashion as beef to reduce STEC, but reports of the efficacy of these interventions are not available.

Contamination of milk with pathogens, including STEC, mainly occurs during milking or via milking equipment, milking personnel, and from the farm environment. Thus, factors affecting the carriage of STEC in live animals and those practices surrounding milking hygiene can reduce, but not assure the absence of contamination of raw milk.

The efficacy of the interventions against STEC during the production of raw milk and raw milk cheeses varied greatly depending on the animal origin of the raw milk, manufacturing practices, the scale of production, and the microbial load. Temperature control and hygiene during milking, storage and transportation can significantly affect the microbiological safety of raw milk prior to processing, packaging and sale of milk intended for drinking or for manufacturing of raw milk cheeses. Although these interventions can mitigate the growth of *E. coli* and other indicator organisms, the degree of support in the evidence for these interventions and the control of STEC ranged from low to medium (Section 2.5).

Apart from pasteurization, which is very effective, several technologies have been evaluated to mitigate the presence of STEC in raw milk. Bacteriophages specific to *E. coli* and STEC have shown some reductions in STEC during refrigeration storage of raw milk (Section 5.1.5). The effect of adding bacteriophage to control *E. coli* during milk fermentation in the making of cheeses has also been examined with varying results depending on the STEC serovar. The degree of support in the evidence of bacteriophage to specifically control for STEC was evaluated as low (Section 5.2.3). Gamma or eBeam irradiation are very effective at reducing bacterial levels in milk and on cheese surfaces, yet off-flavors are often reported. The degree of support for the evidence was rated as medium (Section 5.3.2). A summary of processing and post-processing control measures for STEC in raw milk and raw milk cheese and their degree of support rating (high, medium, low), based on scientific evidence, is available in Annex 4.

The implementation of monitoring plans at the farm level to measure the impact of STEC prevalence is considered impractical, although sampling and testing of beef and raw milk products are a means to verify that food safety program are successful. Because STEC are often present only at low levels in foods, culture enrichment of food samples is a critical step in detecting STEC in meat, dairy and other foods. Since STEC testing is complex, the quantitative detection of non-type specific (NTS) *E. coli* has been proposed as an alternative hygienic indicator during processing and post-processing stages, although it is not an absolute estimate of STEC levels.

The use of molecular techniques, such as PCR, that target STEC virulence genes are highly sensitive and specific for STEC detection but presumptive results must be confirmed by traditional culture-based methods or by immunomagnetic separation (IMS). Methods are needed that enable the efficient and specific isolation of STEC O157:H7 and non-O157 STEC.

The expert committee also discussed some of the limitations and gaps regarding the available data. In-plant scientific evaluations of interventions and treatments to control STEC throughout raw beef, raw milk and raw milk cheese production are frequently prohibited due to health risks associated with the potential introduction of pathogens into the food supply and the cost associated with testing large number of samples required for detecting STEC in food matrices. Consequently, surrogate bacteria, such as NTS *E. coli*, are used as substitutes and the results extrapolated, meaning that evidence of intervention effects specifically for STEC may not be available currently or in the future. Therefore, there is doubt and uncertainty as to whether the detection and reduction levels observed in surrogate studies are truly representative of STEC or of commercial production and processing.

Many studies focused on the impact of an individual control measure at a specific stage in the food chain, rather than in the context a total food chain or of the safety of the food available to the consumer. Many food businesses have implemented multiple control measures concurrently or sequentially on farms and in processing facilities, but the overall efficacy of multiple "hurdles" in the total chain remains difficult to quantify

It was recognized that with advances in analytical methods, including increasing use of molecular tools, the evaluation of evidence concerning some STEC control measures and interventions may need to be revised in the future.

Background
and approach

Shiga toxin-producing *Escherichia coli* (STEC) are an important cause of foodborne disease. Infections can result in a wide range of disease symptoms from mild intestinal discomfort and hemorrhagic diarrhea to severe conditions including haemolytic uremic syndrome (HUS), end-stage renal disease and death. In its report on the global burden of foodborne disease, WHO estimated that in 2010 foodborne STEC caused more than 1.2 million illnesses, 128 deaths, and nearly 13 000 Disability Adjusted Life Years (DALYs) (WHO, 2015).

The Codex Committee on Food Hygiene (CCFH) has highlighted the importance of STEC in foods since its 32nd Session in 1999, when it prioritized their presence in beef and sprouts as significant public health problems in Member countries (FAO and WHO, 2000). Following a request from the 47th Session in November 2015 (FAO and WHO, 2016), the FAO and WHO published the report *Shiga toxin-producing Escherichia coli (STEC) and food: attribution, characterization and monitoring* in 2018 (FAO and WHO, 2018). As part of the 50th session of CCFH in November 2018, the FAO/WHO further updated the committee with additional information on STEC that were subsequently published in the report *Attributing illness caused by Shiga toxin-producing Escherichia coli (STEC) to specific foods* (FAO and WHO, 2019a). The Codex Alimentarius Commission (CAC) at the 42nd Session, July 2019, approved new work on the development of guidelines for the control of STEC in beef, raw milk and cheese produced from raw milk, leafy greens and sprouts (FAO and WHO, 2019b). To support this work, the Joint FAO/WHO Expert Meeting on Shiga toxin-producing *Escherichia coli* (STEC) associated with Meat and Dairy Products was convened virtually from 1 to 26 June 2020 to review relevant measures for pre- and post-harvest control of STEC in animals and foods of animal origins.

Although STEC can be isolated from a variety of food production animals, they are most commonly associated with ruminants from which we derive meat and milk. Because of the widespread and diverse nature of ruminant-derived food production, coupled with the near ubiquity of STEC worldwide, there is no single definitive solution to STEC risk control that will work alone or in all situations. Instead, the introduction of multiple interventions applied in sequence as a "multiple-hurdle scheme" to reduce STEC at several points throughout the food chain (including processing, transport and handling) will be most effective. It is important to note that complementary approaches to reduce STEC must be selected, however STEC control strategies may impact product quantity, quality or production efficiency. This impact must not unduly impact stable access to high quality and safe dietary protein for the world's population. When deciding on interventions to implement, it is necessary to consider regulatory status for both local and export markets. It is also essential to consider the required purpose, cost, space availability and infrastructure of the food processing facility, in addition to environmental impact factors, such as waste and effluent disposal (Gagaoua et al., 2022).

1.1. SCOPE AND OBJECTIVES

Potential STEC intervention strategies can be applied at various stages throughout the food production-to-consumption continuum. For the purposes of discussion, the expert committee has presented three general categories or stages of food production and processing, whilst recognizing that the diversity and complexity of value chains vary considerably depending upon scale, geography and specific products. Hence certain strategies may be classified by others as being part of a different or of multiple categories.

Primary Production control strategies address the stages of animal production until the time of arrival at the slaughter site; and for milk, until milk is stored prior to pasteurization, or for use in raw milk cheese making, or it is further distributed as raw milk. Control strategies during this phase include farm management activities, waste disposal, feed production, milking procedures, and transport prior to slaughter/pasteurization and processing.

Processing control strategies include interventions applied at lairage and animal handling facilities at slaughter operations, included as well are slaughter and dressing processes through to carcass chilling and further fabrication. Treatment of fluid drinking milk intended to be consumed without pasteurization and all steps in the production of raw milk cheeses are included as potential processing controls.

Post-processing control strategies are focused upon raw meat and raw milk (both fluid milk and products) following initial processing through to consumer handling. This stage includes controls applied following carcass production, chilling of carcasses, and those involved in butchering the meat into primal, subprimal and fabricated cuts, packaging, storage, and transport of meats prior to retail or consumer handling. Post-processing controls for raw milk and raw-milk cheeses include interventions and handling processes applied to products (storage, maturation, packaging transport) prior to distribution to wholesalers, retailers or consumers.

During the meeting, the expert committee examined each of these general categories of process steps separately, yet they are clearly interconnected as the production of food is a continuum. Reductions in STEC populations achieved during primary animal production may be nullified by subsequent improper processing and post-processing procedures. Furthermore, there has been inconsistent evidence of a linkage between STEC prevalence in live cattle and contamination of carcasses or finished beef or milk products.

1.2. SYSTEMATIC APPROACH

Interventions and strategies to control STEC were considered when there was evidence from either laboratory, pilot plant or commercial scale studies that demonstrated a reduction in the prevalence or concentration of STEC, non-type specific (NTS) *E. coli*, faecal microorganisms or total microbial load. Moreover, good agricultural practices (GAP) and/or good hygiene practices (GHP) considered to reduce or control the transmission of microorganisms were also included even if there was not specific evidence of their effect on STEC. The quality of evidentiary support varied greatly with study design, scale (laboratory, pilot or full-scale commercial production) and the analytical methods used.

The following sections and the tables in Annex 1, 2, 3 and 4 summarize the interventions evaluated for the specific control of STEC at various critical processing points throughout the beef and dairy value chains. They include examples of research and data from relevant publications. Although the outcomes considered included reduction in prevalence or concentrations of NTS *E. coli* and/or STEC, the scientific evidence was evaluated as to the degree of support based on the interventions ability to specifically control STEC. A scale consisting of the following categories and interpretation was used to qualify the degree of support:

- **Low**: negligible evidence for efficacy of the intervention in reducing STEC presence under field or laboratory conditions and/or documented detrimental effect on animal or derived food product are evident. No obvious pathway or likelihood of immediate adoption.

- **Medium**: published evidence of efficacy of the intervention in reducing STEC presence through research trials and/or supported by laboratory or field experiments and no detrimental effect on animal or derived food product are evident. However, there are concerns about practicality and implementation in the field.
- **High**: clear and compelling evidence that the intervention, management practice, or procedure reduces the presence of STEC at the application point in the production chain. No detrimental effect on animal or derived food product are evident and widespread application is practical.

Many of the practices evalutated at the different stage in the production chain are considered as GAP or GHP and have been scientifically proven to minimizing the potential for microbial contamination, including STEC. Although these well-established best practices have a high degree of support, the purpose of this meeting was to evaluate the scientific evidence and make recommendations specifically regarding their ability to control for STEC.

1.3. BEEF AND DAIRY

1.3.1. Primary production

Primary production control measures for STEC were identified using a variety of laboratory, experimental and epidemiological approaches. Although laboratory studies, and to a lesser extent, animal and research-farm/feedlot experiments, provide opportunity for testing hypotheses under controlled conditions, the extent to which results from these studies can be extrapolated and reasonably expected to be repeatable under real life and commercial settings is often limited. Results from epidemiological studies or ongoing monitoring of animals may provide more realistic data on the effectiveness of interventions under certain field conditions, but such studies are more difficult to control and replicate because of biological variability and differences in production practices, environments and species.

The reporting of well-designed studies evaluating the impact of individual strategies on farms, as well as those strategies applied in sequence to assess potential synergies, will provide greater confidence on the strength of the associations between the interventions and the outcomes. Moreover, primary production control strategy assessments were frequently based strictly on microbiological criteria, with other important metrics such as animal health, well-being, toxicology, and economic impact seldomly reported.

Early microbiological assessments of intervention effectiveness in live animals typically focused on the prevalence of STEC (percentage of animals or farms that are STEC positive) in the study population. Over time some researchers, but not all, included the concentration of STEC present in the specimens as an important outcome variable. Although both prevalence and excretion concentration are important for determining the overall risk of STEC transmission to raw beef or milk, much of the available literature lacks both of these variables, thus limiting the potential for cross study comparisons and understanding the fullest impacts of the interventions and their applicability to the production environment. Thus, both prevalence and concentrations of STEC present are important factors for the determination of intervention efficacy

1.3.2. Beef processing and post-processing

Challenge studies using STEC have been performed, but only a few studies investigated the impact of both low and high inoculum and various strains and serogroups were used in these studies. A published meta-analysis on the effectiveness of interventions for the control of *E. coli* contamination in cattle processing plants found that initial microbial concentrations were significant predictors of intervention effectiveness (Zhilyaev *et al.*, 2017). These results demonstrate that interventions become less effective as *E. coli* counts decrease, even when publications when results with questionable detection limits were excluded.

For post-processing interventions, the experimental approaches were associated with many limitations:

- Many studies used a high starting inoculum concentration of one or more STEC strains. Few studies examined the impact of low and high inoculum levels. Low inoculum levels are likely to be more representative of natural contamination cases.
- Most studies were conducted using a single or mixed strain of STEC O157:H7. A few studies used non-O157 STEC strains, but most of these studies presumed that STEC O157:H7 and non-O157 behaved similarly in those environments.
- Some studies suggested little variation in intervention impacts between different strains, but strain to strain variations have been reported with other interventions, such as bacteriophage treatments.
- The $\log_{10}$ reductions in STEC concentrations reported may not include stressed or injured cells which are unable to recover and be detected when selective media are used for enumeration. Nevertheless, injured cells are viable and may recover under favourable circumstances.
- The study conditions used in some of the experiments may not reflect actual production practices, for example, temperatures and durations of product storage.

- For interventions where there are many experimental studies (e.g. organic acids), meta-analysis should be used to look at the scale of STEC reduction data in more detail.
- Other concerns include the efficiency of methods used in the recovery of STEC cells from different types of beef products, including whether current techniques are capable of recovering all survivors and as such are log_{10} reductions potentially overestimated, and the recovery media used for different meat product/matrices need to be examined and clarified to understand the impact of viable but not culturable (VBNC) STEC cells in these studies.

1.3.3. Dairy processing and post-processing

For this report, rather than being exhaustive, the expert committee's appraisal of the body of scientific evidence focused on interventions to reduce STEC prevalence and concentration in raw milk and raw milk cheeses, predominantly of bovine and caprine origin. Similarly, most of the interventions were evaluated in challenge studies in the laboratory or implemented in pilot plants rather than under commercial, production-scale conditions. Lastly, it was noted that the efficacy of interventions against STEC in raw milk cheeses varied greatly depending on the animal species of origin of the milk, manufacturing practices, the scale of production, the baseline microbial load and composition of the raw materials, and the STEC serotype.

References

FAO & WHO. 2000. Report of the thirty-second session of the Codex Committee on Food Hygiene. Codex Alimentarius Commission. Twenty-fourth session, Geneva, Swtizerland, 2–7 July 2001. Cited 31 March 2022. www.fao.org/fao-who-codexalimentarius/sh-proxy/en/?lnk=1&url=https%253A%252F%252Fworkspace.fao.org%252Fsites%252Fcodex%252FMeetings%252FCX-712-32%252FAl01_13e.pdf

FAO & WHO. 2019a. Attributing illness caused by Shiga toxin-producing *Escherichia coli* (STEC) to specific foods: Report. *Microbiological Risk Assessment Series* No. 32. Rome, FAO. www.fao.org/3/ca5758en/ca5758en.pdf

FAO & WHO. 2019b. Report of the fiftieth session of the Codex Committee on Food Hygiene. Codex Alimentarius. Forty-second session, CICG, Geneva, 7–12 July 2019. Cited 31 March 2022. www.fao.org/fao-who-codexalimentarius/sh-proxy/en/?lnk=1&url=https%253A%252F%252Fworkspace.fao.org%252Fsites%252Fcodex%252FMeetings%252FCX-712-50%252FReport%252FREP19_FHe.pdf

FAO & WHO. 2018. Shiga toxin-producing *Escherichia coli* (STEC) and food: attribution, characterization, and monitoring. *Microbiological Risk Assessment Series* No. 31. Rome, FAO. www.fao.org/3/ca0032en/ca0032en.pdf

FAO & WHO. 2016. Report of the forty-seventh session of the Codex Committee on Food Hygiene. Codex Alimentarius. Thirty-ninth session, Rome, Italy, 27 June to 01 July 2016. Cited 31 March 2022. www.fao.org/fao-who-codexalimentarius/sh-proxy/en/?lnk=1&url=https%253A%252F%252Fworkspace.fao.org%252Fsites%252Fcodex%252FMeetings%252FCX-712-47%252FReport%252FREP16_FHe.pdf

FAO & WHO. 2014. Report of the forty-fifth session of the Codex Committee on Food Hygiene. Codex Alimentarius. Thirty-seventh session, Geveva, Switzerland, 14–18 July 2014. Cited 31 March 2022. https://www.fao.org/fao-who-codexalimentarius/sh-proxy/en/?lnk=1&url=https%253A%252F%252Fworkspace.fao.org%252Fsites%252Fcodex%252FMeetings%252FCX-712-45%252FREP14_FHe.pdf

Gagaoua, M., Duffy, G., Alvarez, C., Burgess, C. M., Hamill, R., Crofton, E., Botinestean, C., Ferragina, A., Cafferky, J., Mullen, A. M., Troy, D. 2022. Current research and emerging tools to improve fresh red meat quality. *Irish Journal of Agricultural and Food Research.* doi: 10.15212/ijafr-2020-0141

WHO. 2015. WHO Estimates of the global burden of foodborne diseases. Geneva, WHO. https://apps.who.int/iris/bitstream/handle/10665/199350/9789241565165_eng.pdf?sequence=1

Zhilyaev, S., Cadavez, V., Gonzales-Barron, U., Phetxumphou, K. & Gallagher, D. 2017. Meta analysis on the effect of interventions used in cattle processing plants to reduce *Escherichia coli* contamination. *Food Research International*, 93: 16–25. doi: 10.1016/j.foodres.2017.01.005

Primary production control strategies for STEC in beef and dairy

Primary production controls include strategies that can be used on small and large farm operations for food animal species raised for meat and dairy production. There are many potential control points during the life cycle of food animals where STEC carriage and/or prevalence can be reduced. Because of the diversity of animal production systems world-wide, no single factor can be universally applied to control STEC in every circumstance and condition. Moreover, many control methods discussed herein are broadly impactful, such as the application of GAP, to reduce the spread of many pathogens, including STEC, but are not specifically targeted against STEC.

While cattle are not the only ruminant species to carry STEC, beef products or cross-contamination from cattle husbandry (or production) are more frequently associated with human foodborne STEC illnesses. Moreover, in addition to contributions of dairy cattle to human STEC illness via consumption of contaminated milk products, the majority of dairy cattle at the end of their productive lives, also enter the beef production chain; therefore, this report has focused primarily on beef and dairy cattle. Nevertheless, many of the concepts and principles for STEC control may also be applicable for other food animals, especially other ruminants (Section 6). Because of the paucity of literature on the control of STEC in animals other than *Bos taurus* (and *Bos taurus indicus*), this section addresses control of STEC in cattle.

Many of the cattle management strategies discussed fall into GAP or GHP, but it is important to note that these practices can have a significant impact on the risk of STEC carriage and transmission to the food supply. Many of the practices reported herein are not necessarily specific to STEC but can also be useful controls for other pathogenic foodborne bacteria such as non-typhoidal *Salmonella*, *Campylobacter*, and *Listeria*. However, the efficacy of any intervention against STEC should not be construed to imply their effectiveness against other foodborne pathogenic bacteria, or vice versa.

A summary of primary production control measures for STEC in cattle and their degree of support rating (high, medium, low), based on scientific evidence, is available in Annex 1.

2.1. ANIMAL FACTORS

Although STEC are commonly found in ruminants, specific animal factors (e.g. host breed, genetics) can influence host animal carriage and excretion of STEC. Animal-specific differences have not been exploited for the development of intervention strategies per se, yet they represent areas for future investigation (e.g. specific targeting of host epithelial cells via the intimin protein vaccine development) to reduce STEC excretion and prevalence.

Super-shedding cattle are defined as those that are excreting more than 4 $\log_{10}$ CFU/g of STEC O157:H7 of faeces (Munns *et al.*, 2014; Munns *et al.*, 2015; Castro *et al.*, 2017). A substantial body of evidence demonstrated that super-shedding events are responsible for the majority of STEC transmission among cattle and is the largest contributor of STEC spread and a threat to food safety. However, the detection of a super-shedding event is dependent on the timing of faecal sampling of an individual animal, possibly a reflection of the natural course of infection in cattle following exposure. To date, efforts to identify distinct genomic markers in the host related to super-shedding has been largely proven unsuccessful. There is evidence that certain strains of STEC O157:H7 are more likely to be associated with super-shedding, but this relationship is not conclusive (Munns *et al.*, 2015). One hypothesis that has been posited is that super-shedding results from STEC O157:H7 biofilms within the lumen of the rectum being sloughed from the intestinal epithelium and entering faecal matter (Munns *et al.*, 2014). This event would account for the high levels of STEC O157:H7 in faeces and the intermittent nature of the phenomena.

2.1.1. Cattle genetics

There is a growing body of evidence indicating that host genetics impact both the phylogenetic diversity and the relative abundance of populations of the intestinal microbiome. Factors such as innate and acquired immunity as well as other host-microbiome communication channels may influence the establishment of STEC within the host and subsequently, STEC excretion (Wang *et al.*, 2016). Suppression of some of the genes involved in immune responses within the epithelium of the recto-anal junction has been observed in cattle that tend to shed more STEC O157:H7 (Jeon *et al.*, 2013). However, currently it is not clear if this reflects a response controlled by host genetics or one that is induced as a result of interactions between STEC O157:H7 and the host epithelium. We are unaware of studies specifically designed to assess the impact of cattle breed on STEC excretion, but it is clear that both beef and dairy cattle can readily become colonized with these pathogens (Jeon *et al.*, 2013; Riley *et al.*, 2003). Production conditions, animal management, diet composition and STEC genomics are likely more important and impactful than cattle host genetics in determining the likelihood of STEC excretion.

> *The degree of support for the use of host animal genetics as an intervention specifically for the control of STEC in cattle herds was low.*

2.1.2. Cattle intestinal microbiome

The large microbial ecosystem that occupies the gastrointestinal tract of cattle is a vast biochemical/enzymatic/nutrient reservoir that can impact STEC carriage. There is evidence that the phylogenetic composition of this microbial population within the lower gastrointestinal tract varies between super-shedding and non-shedding cattle, with a greater microbial diversity found in non-shedders (Zaheer *et al.*, 2017; Xu *et al.*, 2014). However, distinct members or catabolic niches within the microbial community that confer resistance (or susceptibility) to STEC colonization have not been identified. Furthermore, end products of the gastrointestinal microbial fermentation that enhance or prevent STEC colonization remain unknown. Prebiotics and probiotics may also alter the excretion of STEC through changes in the species composition of the gastrointestinal microbiome. Shifts in the host microbiome may alter nutrient availability for the growth of STEC or metabolic end products from microbial fermentation that inhibit the growth and establishment of STEC within the intestinal tract (VanKessel *et al.*, 2002). However, it is apparent that changes in the microbial population of the gut can provide opportunities for pathogens to colonise the gut, or to be excluded from the complex population.

The degree of support for the modification of the gastrointestinal microbiome as an intervention specifically for the control of STEC in cattle heards was low.

2.1.3. Cattle demography (age, sex, production status)

Age of the animal, and the type of cattle have been identified as risk factors for STEC excretion. Younger animals have greater susceptibility to STEC colonization and STEC faecal prevalence is higher in young milk-fed animals as compared to mature adult beef and dairy cows. Differences in colonization are thought to be related to changes in the animal genetic and physiological factors, diet, management and environmental factors that occur as cattle approach physiologic maturity (Ekong, Sanderson and Cernicchiaro, 2015).

Grouping cattle into age specific groups is a GAP that minimizes disease spread and prevents exclusion and bullying by dominant animals. Calves shed STEC O157:H7 more frequently than do older cattle, so keeping young cattle in the same groups throughout rearing without introducing new animals is recommended to reduce STEC O157:H7 prevalence (Sanderson *et al.*, 2006; Ellis-Iversen *et al.*, 2008). Other factors such as feeding colostrum from the calf's dam instead of from pooled sources, a decrease in serum IgG concentration and a high temperature-humidity index in the pens, were identified as increasing the likelihood for the presence of STEC O157:H7 in pre-weaned calves. These findings supported the idea that immunity and overall health of the calf gastrointestinal tract are important factors that may influence the faecal presence of STEC O157:H7 (Stenkamp-Strahm *et al.*, 2018). To prevent the spread of animal disease and enhance animal health, the housing of young dairy calves and veal calves individually or in small groups is common practice and can contribute to a reduction in the transmission of STEC (Bosilevac *et al.*, 2017).

Faecal prevalence of STEC O157:H7 may also be impacted by the production stage of the cattle. However, it should be noted that eventually many dairy animals enter the meat supply at the end of their dairy production stage. The prevalence of STEC serogroups often associated with severe infections in people, such as O157, O26, O111, O103, O121, O45, and O145 was significantly higher in younger animals (veal, young beef and dairy) than in adults (beef and dairy) (Bibbal *et al.*, 2015; Mellor *et al.*, 2016), but the basis for this difference remains to be elucidated. If possible, separation of calves from adult cows and replacement dairy heifers, and lowering the density of cattle in pens, may reduce STEC prevalence by reducing opportunities for animal-to-animal horizontal spread (Ekong, Sanderson and Cernicchiaro, 2015). It has been shown that there was an increased probability of finding a STEC excreting animal where there are larger numbers of finishing cattle, or a farm being classified as a dairy unit that also stocked beef animals (Gunn *et al.*, 2007; Synge *et al.*, 2003).

The impact of stage of lactation on STEC excretion is inconsistent (Edrington *et al.*, 2004). During the first 30 days (approximately) after calving, dairy cows rapidly increase milk production capacity, which consumes a great deal of dietary energy. During this transition period, many high milk-producing dairy cows cannot consume enough feed to meet their energy demands, resulting in severe negative energy balance, which in turn is linked with many metabolic disorders and is associated with a 30 percent increase in the risk of STEC O157:H7 prevalence and excretion in dairy cows (Venegas-Vargas *et al.*, 2016). There is currently no evidence indicating that there is an effective intervention strategy or control point linked with reducing the effects of the negative energy balance. There are no specific recommendations to address negative energy balance as a hazard-based intervention for the control of STEC in cattle herds, except for following GAP measures to improve animal immunity and overall animal health, which could minimize STEC spread.

> *Grouping of cattle by age and production status is a GAP; however, the degree of support for the use of cattle grouping based on demographic characteristics as an intervention specifically for the control of STEC in cattle herds was medium.*

2.2. ENVIRONMENTAL FACTORS

2.2.1. Biosecurity

While cattle, sheep and goats are primary reservoirs of STEC, other ruminants as well as monogastric animals can carry STEC. These other animal species can spread STEC between animals on a farm (Bolton, O'Neill and Fanning, 2011), from farm-to-farm, and in the case of birds, over long distances (Cernicchiaro *et al.*, 2012).

Mixing of sheep with cattle has been shown to increase the risk of cattle shedding STEC (Stacey *et al.*, 2007), and a positive correlation was found between cattle shedding STEC and sheep density (Strachan, Fenlon and Ogden, 2001).

Studies have found that other animals (rodents, insects, birds and boars) can carry STEC at least transiently, which may increase the risk of STEC shedding by cattle (Ahmad, Nagaraja and Zurek, 2007; Branham *et al.*, 2005; Cernicchiaro *et al.*, 2012; Cizek *et al.*, 1999; French *et al.*, 2010; Hancock *et al.*, 1998; Talley *et al.*, 2009; Rice *et al.*, 2003; Sánchez *et al.*, 2010; Wetzel and LeJeune, 2006). The presence of dogs, pigs, or wild geese on the farm have also been associated with an increased risk of STEC O157:H7 shedding (Gunn *et al.*, 2007; Synge *et al.*, 2003). These vectors can transfer STEC (and other pathogens) to non-infected groups of cattle within a farm, between farms and even over long distances.

Another important consideration is open versus closed herds. Maintaining a stable herd population and not moving cattle between herds is a GAP and contributed to a 48 percent reduction in STEC in a randomised control trial in England and Wales (Ellis-Iversen *et al.*, 2008).

> *On-farm biosecurity is a GAP, and the degree of support for farm biosecurity as an intervention specifically for the control of STEC in cattle herds was medium to high.*

2.2.2. Animal density

Higher animal density has been linked with an increased risk of carriage of some STEC, including STEC O157:H7, and may also play a role in the horizontal spread of STEC O157:H7 (Frank *et al.*, 2008; Gunn *et al.*, 2007; Vidovic and Korber, 2006), since densely packed animals have a greater chance of contamination via faecal spread in feed and water supplies and on hides. Stocking density of cattle is especially important when super-shedding animals are present as an increase in physical contact between animals results in higher environmental concentrations of STEC. Animal density also plays a role in the recirculation of STEC population within a herd by increasing individual exposures to faeces from other ruminants.

> *The degree of support for reducing animal density on-farm as an intervention specifically for the control of STEC in cattle herds was medium.*

2.2.3. Environmental hygiene

Sanitary conditions are important for healthy animal production and are regarded as best practices for animal production. Proper sanitation of facilities and disposal of manure are important for interrupting the faecal-oral transmission of STEC and other similarly transmitted bacterial pathogens (Garber *et al.*, 1999).

Contaminated materials on pen floor also represent a substantial source of STEC O157:H7 that can drive population dynamics. Contamination spread may be subsequently amplified by flushing alleyways with water intended to remove manure or by rainfall events in open pens. Wet conditions result in muddy pen floors which can impact animal performance, welfare, and can increase STEC survival.

Pen floor conditions can impact STEC colonization of the hides of feedlot cattle, and muddy pens also increased the chance of STEC excretion (Smith *et al.*, 2001). Minimizing mud and cleaning pen surfaces between groups of cattle are part of GHP that are not specific for STEC but are thought to reduce STEC prevalence in and on cattle.

During food animal production, the animals must be restrained to ensure GAP (e.g. vaccination). Often, the use of a restraint system of chutes and squeezes will have physical contact with the animal hide (Mather *et al.*, 2007). As a result, faecal material on the hide of one animal can be transferred horizontally between animals in the same facility and may spread through the entire herd or into receiving pens (Cobbold and Desmarchelier, 2002). Cleaning and disinfection of the common handling area and facilities should reduce risks of horizontal transmission of STEC and other pathogens, though this has not been fully substantiated.

> *Adequate facility, bedding material and pen floor maintenance and hygiene are GAP, and the degree of support for these to be used as an intervention specifically for the control of STEC in cattle herds was medium.*

2.2.4. Manure management issues

If not properly treated or contained, manure can serve as a source of STEC dissemination on the farm and the broader environment. Manure that is not adequately composted prior to its application to either pasture or cultivated fields as a soil amendment, can contain pathogens that can be transmitted back to the cattle herd via farm-grown forages or silages. STEC in manure can enter either surface or ground water and can pose a risk of contamination of drinking water, wells, ponds or irrigation water supplies (Blaustein *et al.*, 2015). Risks associated with manure application in agriculture can be reduced by secondary manure treatments such as bio-digestion, composting, stock piling or desiccation (Ongeng *et al.*, 2015). As STEC are found to be persistent in cattle faeces, management of manure and slurries on the farm can influence spread of the pathogens (Duffy, 2003; Callaway *et al.*, 2013). Vegetation strips may reduce the flow of STEC with surface water during rainfall events, but if the cattle are allowed to graze on recently manured pastures, the possibility of faecal-oral transmission of STEC is increased.

> *Adequate manure managment is a GAP; however, the degree of support for manure management to be used as an intervention specifically for the control of STEC in cattle herds was low.*

2.2.5. Seasonal variability and temperature

The seasonality of STEC, especially O157:H7, excretion among cattle (e.g. "summer peak") and its correlation with the incidence of human illness is well documented (Money *et al.*, 2010; Ekong, Sanderson and Cernicchiaro, 2015). While we still do not know the cause of STEC seasonality, there are multiple potential reasons, several of which are linked to summer months and higher temperatures (e.g. increased growth of *E. coli* and STEC in water troughs and increased survival in excreted faeces), proliferation of other vectors (e.g. protozoa), longer days, greater animal congregation in shadows and at feeding sites at cooler times of day, and to a lesser extent, drinking behaviours (Bach, Stanford and McAllister, 2005b; Gautam *et al.*, 2011; Besser *et al.*, 2014; Dawson *et al.*, 2018).

Heat stress in cattle has shown limited impact on STEC excretion (Brown-Brandl *et al.*, 2009; Edrington *et al.*, 2004). Alleviating heat stress is an animal welfare issue frequently implemented in GAP. Use of sprinklers on animals to alleviate heat stress demonstrated no impact on STEC O157:H7 populations (Morrow *et al.*, 2005). However, sprinklers can increase the presence of mud in pens, which can increase the survival of STEC and can result in increased amounts of STEC contaminated mud and faeces contaminating an animal's coat (Edrington *et al.*, 2009a).

Alleviating heat stress in cattle is a GAP; however, the degree of support for using sprinklers as an intervention specifically for the control of STEC in cattle herds was low.

2.3. WATER AND FEED MANAGEMENT STRATEGIES

2.3.1. Drinking water quality and hygiene

Drinking water quality is assumed to be part of GAP in cattle rearing and can have a profound impact on animal production. However, microbiological quality of water itself has not been shown to impact STEC populations or prevalence in cattle.

Cleaning of water troughs has ben suggested to reduce the load of NTS *E. coli* and other bacterial populations in the water troughs, which theoretically would reduce dissemination of STEC in cattle production. However, direct evidence on the efficacy of trough cleaning has not been demonstrated. Keeping water levels high in water troughs was suggested to reduce concentration of STEC O157:H7 via simple dilution, and the ratio of cattle to water volume impacts the concentrations of faecal contamination and the horizontal spread of faecal microbes, including STEC (Beauvais *et al.*, 2018).

Quality drinking water is a GAP; however, the degree of support for clean drinking water as an intervention specifically for the control of STEC in cattle herds was low.

2.3.2. Drinking water treatment

Chlorine is an effective disinfectant that is used extensively as an antimicrobial treatment in potable water and cleaning. It has been broadly assumed by many cattle producers that the addition of chlorine to drinking water could reduce STEC loads in water troughs, however this has not been directly demonstrated. Furthermore, the antimicrobial activity of chlorine is reduced when there are high levels of organic matter in the water and also when exposed to UV from sunlight (LeJeune, Besser and Hancock, 2001). Contamination of water troughs with feed from the oral cavity of cattle during drinking or faecal matter can reduce the effectiveness of chlorination. Chlorine can be used for water trough cleaning and decontamination, but its effectiveness is found to be short term as these surfaces rapidly become re-contaminated shortly after cleaning (Smith *et al.*, 2002). Electrolyzed-oxidized (EO) water administered to drinking water was more effective in reducing STEC populations than chlorinated water but EO is also subject to inactivation by organic matter and UV light (Bosilevac *et al.*, 2005; Stevenson *et al.*, 2004). Drinking water is only one of many possible transmission pathways of STEC in the production environment and it is likely that direct animal-to-animal or faecal-to-animal contact are more significant mechanisms of transmission (LeJeune *et al.*, 2004).

The degree of support for the treatment of drinking water as an intervention specifically for the control of STEC in cattle herds was low.

2.3.3. Diet composition, feeding strategies and feed hygiene

Using cattle feedstuffs that are free of pathogens is a GAP and subject to regulations by most countries. However, feeds may be contaminated by microbes from a variety of sources, ranging from the time of productions, processing, transportation and storage up until the feed is consumed. It has been suggested that feed troughs are potential reservoirs of STEC contamination; however, there is limited evidentiary support for this or for it to be an effective STEC control point. Additionally, there are numerous logistical hurdles in the production operations that all together, makes consistent feed trough hygiene difficult. As a result, this strategy has limited efficacy.

The type of feed ingredients and the way that ingredients are processed prior to feeding has also been proposed to impact the proliferation or suppression of STEC populations in the ruminant gastrointestinal tract, possibly through modifications of gut microbiome, the presence and proportion of specific metabolites (e.g. volatile or short chain fatty acids) and faecal pH.

Forage: Concentrate ratio

Escherichia coli populations (both NTS and STEC O157:H7) are present in cattle fed forage-based diets and in those fed rations high in grains. Experimental studies on the duration or magnitude of excretion of faecal STEC O157:H7 populations among calves that were fed high grain or high forage (low quality forage) diets, showed variable results (Callaway *et al.*, 2009, 2013).

> *The degree of support for forage feeding, as compared to rations higher in grain or concentrates, as an intervention specifically for the control of STEC in cattle herds was low to medium.*

Dietary shifts from grain to forage

A rapid shift from a high-grain to a higher fibre ration was initially proposed as a mechanism to control STEC (Diez-Gonzalez *et al.*, 1998). However, subsequent experiments produced inconsistent results (Hancock *et al.*, 2000, Hovde *et al.*, 1999, Keen *et al.*, 1999). Also, dietary shifts to high-forage rations are difficult to implement in feedlots and in arid regions as there are many other logistical challenges. Rapid dietary changes or feeding a high-grain rations results in a shift in the microbial population and creates ecological niches which may be exploited by pathogenic organisms for colonization (Jacob, Callaway and Nagaraja, 2009). However, the mechanism of action of diet change on the concentrations of faecal *E. coli* is not fully understood.

> *The degree of support for a rapid shift from a grain-based to forage-based rations as an intervention specifically for the control of STEC in cattle herds was low.*

Grain type

For many years, the rumen was thought to be absent of *E. coli* populations due to the inhibitory effects of short chain fatty acids (SCFA) (Wolin, 1969). However, studies over the past 20 years have found that STEC are frequently isolated from the rumen of cattle that were fed forages and grains. Cereal grains, due to their high starch content, are often included in the diet of ruminant animals to increase their energy density. Most dietary starch is fermented rapidly in the fore-gut (rumen) of cattle and results in the production of SCFA and lactic acid; however, not all starch in the rumen is fermented and some passes to the hindgut (cecum and colon) where it undergoes a secondary fermentation.

The rate of fermentation of cereal grains varies with grain type and the physical grain processing method used, as cereal grains differ in their composition and starch availability to the microbial population. Starch fermentation results in profound changes in the rumen and the hindgut, including decreases in pH, forage (cellulose and hemicellulose) degradation, as well as decreases in ruminal microbial diversity. The rate of starch fermentation impacts ruminal SCFA production and pH, which in turn, alters the composition of the ruminal microbial ecosystem and the ability for pathogens to survive or colonize ruminants (Berg et al., 2004).

Barley is a cereal grain which is more rapidly fermented in the rumen than corn, so little barley starch passes to the hindgut. Feeding barley to cattle increased the incidence of excretion and the concentration of STEC O157:H7 by <0.5 $\log_{10}$ CFU/g in faeces compared to feeding corn. Also, feeding barley to cattle increased the survival time of STEC O157:H7 (above the limit of detection) in faeces by more than 2 fold when compared to corn (Berg et al., 2004; Bach et al., 2005a). These factors could influence the survival of STEC O157:H7 in the lower digestive tract, where STEC appears to colonize primarily.

> The degree of support for the feeding of diets based on starches with slower fermentation rates as an intervention specifically for the control of STEC in cattle herds was medium.

Grain processing

Grain used in cattle feed can be processed by various methods to improve digestibility and to change the site of starch fermentation to the rumen or to the hindgut. Dry rolling of corn (cracking), instead of steam flaked corn, increased starch flow to the hindgut, and reduced STEC O157:H7 faecal concentrations (Fox et al., 2007). While, steam flaking of corn increased STEC O157:H7 excretion in heifers as compared to feeding whole corn. Faecal starch concentration and pH were not linked to the faecal excretion of STEC O157:H7, yet post-ruminal starch infusion increased NTS E. coli concentrations in the hindgut. The exact mechanisms that the methods of grain processing exert on STEC populations remains unclear, but steam flaking of corn seems to increase risks of STEC excretion (Depenbusch et al., 2008).

> The degree of support for avoiding feed with steam flaked corn grains as an intervention specifically for the control of STEC in cattle herds was medium.

Distiller's grains with solubles

Industrial ethanol production from the fermentation of corn starch results in a by-product known as distiller's or brewer's grains that can be used as animal feed. Distiller's and brewer's grains are low in starch, having been removed in the distillation process, and can be fed to cattle as a low-starch protein source. Distiller's grains (DG) are fed either as wet distiller's grains with solubles (WDGS) or as dry distiller's grains with solubles (DDGS), but inclusion of either of these forms of DGS, or the inclusion of brewer's grain in cattle feed, has been associated with increased faecal prevalence and excretion of STEC O157:H7 (Wells *et al.*, 2011; Berry *et al.*, 2017). The magnitude of effect of DGs on STEC sheding and the survivial of STEC in the faeces of animals fed this feed ingredient may be influenced by the particular characteristics of the DG caused by variability in production conditions.

> *The degree of support for feeding distiller's or brewer's grains at levels below 15 percent of ruminant ration as an intervention specifically for the control of STEC in cattle herds was medium.*

Tannins and essential oils

Tannins and essential oils are known to impact the microbial community composition in the gastrointestinal tracts of ruminants, in a unique fashion related to the mode of action of each of these compounds. Inclusion of tannin-containing components or extracts in the ration can reduce STEC O157:H7 and other STEC serotypes in the faeces of cattle, with phlorotannin-containing air-dried brown seaweed being more effective than terrestrial tannin sources (Braden *et al.*, 2004; Zhou *et al.*, 2018). Feeding citrus products and other essential oils also impacted faecal and ruminal STEC O157:H7 concentrations, but these have only been assessed in very small-scale studies (Callaway *et al.*, 2011).

> *The degree of support for the inclusion of tannins and essential oils in ruminant rations as an intervention specifically for the control of STEC in cattle herds was low to medium.*

2.3.4. Feed additives

Probiotics / Direct fed microbials / Competitive exclusion

The prevalence of faecal STEC O157:H7 excretion in cattle can be reduced by using direct-fed microbials (DFM), such as *Lactobacillus acidophilus* (NP51) and *Propionibacterium freudenreichii* (NP24) (Wisener *et al.*, 2015). Competitive exclusion by probiotic *E. coli* strains orally administered to calves reduced faecal excretion of STEC O26 and O111 and may also represent a promising control measure for reducing other STEC serotypes in cattle (Zhao *et al.*, 2003). To be effective, the component strains in the product must be consistent and these products must be administered at the recommended CFU/g doses in feed. The impact of DFM against STEC is highly specific, thus a positive response in STEC reduction with one probiotic product cannot be necessarily extrapolated to another product.

The degree of support for the use of some DFM as an intervention specifically for the control of STEC in cattle herds was medium.

Bacteriophage

Bacteriophages have been shown to be very effective at killing STEC on food handling surfaces and on the surface of food, and in some cases reducing STEC O157:H7 to undetectable levels (Liu *et al.*, 2015). In primary production steps, however, the efficacy of bacteriophages has been mixed, with either little reduction in STEC numbers or the establishment of a cyclic effect where STEC numbers decline when phage numbers are high, but as phage numbers decreases, the STEC numbers increases (Sabouri *et al.*, 2017). Phage have been used in a commercial product to control STEC O157:H7 on cattle hides prior to slaughter, but the results were marginal (Arthur *et al.*, 2017).

Contact between the phage and the targeted bacteria is essential for the phage to be effective. This is a stoichiometric process that depends on the concentrations of the phage and the host bacterium, as well as the consistency of the matrices that enable adequate mixing of phage and host. Bacteriophages are most effective when large numbers of phage plaque forming units (PFU) can be applied and where contact with the target bacterium can be assured (Wang *et al.*, 2017). Effects of phage may be very broad or may be highly host specific and can be limited to one or a few STEC serotypes. When exposed to phage, STEC can also become resistant to phage either due to the activation of CRISPR/CAS defence mechanisms which prevent phage DNA insertion, or alteration in the host surface receptors which the phage relies on for recognition. Use of phage mixtures has been proposed as means of overcoming phage resistance, but there is also evidence that this approach can select for bacteria that are resistant to multiple phage types. Consistent reductions of STEC by phage treatment are more difficult to achieve in complex matrices and situations such as in the intestinal tract of live animals, where interactions between the phage and the targeted bacterium cannot be assured.

The degree of support for the addition of bacteriophage to feed as an intervention specifically for the control of STEC in cattle herds was low.

Colicins

Colicins are antimicrobial proteins produced by *E. coli* and specifically target *E. coli*, but do not differentiate between NTS *E. coli* and STEC. Colicins has been shown to kill STEC O157:H7 strains *in vitro* and *in vivo* (Callaway *et al.*, 2004; Schulz *et al.*, 2015). *E. coli* colicin was costly to produce and scarce, but colicin genes have been genetically inserted into plants and fungi to allow for scale up in colicin production. Colicin-producing plants and fungi would be considered genetically modified organisms (GMO) for this purpose. Colicins can be added to cattle rations, however the anti-STEC activity of colicin has been shown to be more promising in ground beef and in cleaning processing facility surfaces than in primary production. Colicins are not currently commercially available.

> *The degree of support for the addition of colicins to feed as an intervention specifically for the control of STEC in cattle herds was low.*

Sodium chlorate

Escherichia coli, including STEC can respire anaerobically using the enzyme nitrate reductase, which reduces sodium chlorate to chlorite, a bacteriocidal agent that accumulates intracellularly. In cattle, experimentally, use of sodium chlorate reduced inoculated STEC O157:H7 populations by 2-3 $\log_{10}$/g digesta throughout the gut (Callaway *et al.*, 2002). Chlorate had no impact on meat quality and radiolabelled residue studies showed that it did not accumulate in the tissue. Use of sodium chlorate has also been studied in other animals, but may not be approved by regulatory agencies and is not currently commercially available as a feed additive.

> *The degree of support for the addition of sodium chlorate to feed as an intervention specifically for the control of STEC in cattle herds was medium.*

2.4. VACCINES AND CLINICAL ANTIMICROBIALS

2.4.1. Vaccines

Various vaccines have been designed and tested for preventing colonisation and/or reducing faecal excretion of STEC O157:H7 in cattle. To date, the following vaccine types have been tested: vaccines based on type III secretion proteins (T3SS), vaccines based on siderophores (SRP) and porin proteins, vaccines based on bacterins and bacterial envelopes, vaccines based on flagellin, vaccines based on attenuated *Salmonella enterica* serovar Dublin expressing intimin, vaccines based on Shiga toxin toxoids and vaccines based on outer membrane vesicles

(Smith, 2015; Larzábal, Cataldi and Vilte, 2019; Besser *et al.*, 2014; Fingermann *et al.*, 2018). These vaccines have been tested in feedlot cattle, pregnant cattle, calves and mice. However, only a few vaccines have been tested under production conditions and the duration of immunity after vaccination is unknown because the evaluation period in the feedlot studies has been relatively short (Smith, 2015). Furthermore, two vaccines: the T3SS based vaccine (Bioniche Life Sciences Inc., Belville, Ontario, Canada) and the SRP protein-based vaccine (Epitopix, LLC, Wilmar, Minnesota) have been commercialised (Besser *et al.*, 2014). Both have been tested in the United States of America and were shown to be effective in reducing excretion of STEC O157:H7 in cattle faeces (Snedeker, Campbell and Sargeant, 2012; Varela, Dick and Wilson, 2012). The use of a "cocktail" of different vaccine types has been recommended to enhance vaccine efficacy against STEC O157:H7. Adoption of vaccines by producers is hindered if more than one dose is required to achieve an immune response. At present, neither commercialized vaccine is widely used.

> *The degree of support for the use of vaccines as an intervention specifically for the control of STEC in cattle herds was low to high, depending on formulation considered.*

2.4.2. Clinical antimicrobials

Ruminant animals have a symbiotic relationship with their ruminal and gastrointestinal microbial population, which allows ruminants to thrive on diets that monogastric animals cannot. Unfortunately, this comes at the cost of lower feed efficiency (kg of feed to produce kg of meat/milk). Ionophores, tylosin, chlortetracycline, and oxytetracycline are routinely fed to cattle in some production systems (added to feed and water at subtherapeutic levels) in many countries to improve feed efficiency and improve animal health.

The efficacy of clinical antimicrobials against STEC in the gut of cattle has not been demonstrated *in vivo* or *in vitro* except for the use of neomycin sulfate (described below). However, it is hypothesized that the use of broad spectrum or bacteriostatic and/or bactericidal agents targeted against Gram-negative bacteria, would most probably reduce STEC colonization and excretion. Key considerations contraindicating the use of antimicrobials would include: 1) Some antimicrobials may induce Stx-encoding bacteriophages that are able to transduce Stx-encoding genes and antimicrobial resistance genes to naive *E. coli* thereby contributing to the expansion of the STEC pool as shown in *in vitro* studies (Kimmitt, Harwood and Barer, 2000; Köhler, Karch and Schmidt, 2000); 2) Some antimicrobials may exert selective pressure on intestinal microbiota thereby favouring the survival of antimicrobial resistant STEC (no studies available yet on this issue) and other intestinal microflora;

and 3) Antimicrobials that inhibit Gram-positive bacteria (which are responsible for rapid starch fermentation, and can include opportunistic pathogens) may favour the dominance of Gram-negative bacteria in the gut, including STEC.

The use of antimicrobials to reduce STEC colonization or excretion remains controversial primarily because of concerns with increasing antimicrobial resistance, which may impact public health, globally. No antimicrobial with proven efficacy in reducing STEC in cattle has been demonstrated in scientific studies (Jacob *et al.*, 2008), except neomycin, noted below. Due to valid concerns regarding the use of antimicrobials in animal husbandry, some countries have banned the use of antimicrobials as growth promoters in livestock production.

> *The degree of support for the use of antimicrobials in cattle feed as an intervention specifically for the control of STEC in cattle herds was low to medium.*

Neomycin sulfate

Neomycin sulfate is a commercially available antimicrobial which has been demonstrated to reduce STEC O157:H7 populations in cattle; however, this research study was done on a small scale, so the data is limited and it also raises significant concerns about increasing antimicrobial resistance (Elder *et al.*, 2002).

> *The degree of support for the use of neomycin sulfate as an intervention specifically for the control of STEC in cattle herds was low. The use of neomycin sulfate is not recommended due to potential adverse impacts on public health.*

Ionophores

Ionophores are a class of antimicrobial that inhibit Gram-positive bacteria. Ionophores are not used in human medicine, but have been tested in animals and in beef (and some dairy) cattle feed. They improved feed efficiency by altering the microbial population of the rumen, resulting in a shift in fermentation and end products, including reduced methane production. In theory, inhibition of ruminal Gram-positive bacterial species would benefit Gram-negative species such as STEC. However, this has not been demonstrated in sheep or cattle (Edrington *et al.*, 2003). The percentage of steers excreting enumerable STEC O157:H7 levels was greater in monensin-fed cattle as compared to controls, but feeding ionophores did not change the incidences of STEC excretion in the pens (Hales *et al.*, 2017; McAllister *et al.*, 2006).

> *The degree of support for the use of ionophores as an intervention specifically for the control of STEC in cattle herds was low.*

2.4.3. Beta-agonists / hormones

Ractopamine hydrochloride and zilpaterol hydrochloride are beta-agonists that are fed to finishing cattle in North America for a period of 30–40 days prior to slaughter to enhance weight gain. The use of zilpaterol has been largely discontinued, but ractopamine is still widely used in intensive feedlot cattle production. These additives repartition energy from fat towards muscle and improve feed efficiency. It was hypothesized that this shift in metabolism may increase the "metabolic stress" on the animal and result in increased pathogen excretion or that beta-agonists may impact bacterial quorum sensing and increase virulence gene expression in microbial pathogens. A series of studies were undertaken to investigate the impact of these additives on the excretion of STEC O157:H7 (Wells *et al.*, 2017; Paddock *et al.*, 2011; Edrington *et al.*, 2009b). Although there was variation among the studies, none found that the addition of beta-agonists to the cattle diet resulted in an increase in the excretion of STEC O157:H7. One study even found a reduction in STEC O157:H7 faecal excretion as a result of the administration of ractopamine hydrochloride. At this point, the evidence would indicate that beta-agonists do not have an impact on the excretion of STEC O157:H7 in cattle. The expert committee was unable to find other studies that investigated the impact of other hormones such as bovine somatotropin in dairy cattle or estrogenic or androgenic implants/additives in beef cattle.

The degree of support for the use B-agonists as an intervention specifically for the control of STEC in cattle herds was low.

2.5. DAIRY PRODUCTION SPECIFIC INTERVENTIONS

Contamination of raw milk can occur as a result of infection in the animal prior to milking, from contact with faecal material directly from an animal during milking or via milking equipment, from personnel or from the farm environment. Despite GAP and GHP that are essential to minimize bacterial contamination, several other opportunities for microbial contamination of raw milk occur during processing, thus bacteria cannot be eliminated completely regardless of whether the raw milk is intended for drinking or for use in the production of raw milk cheeses (Jayarao and Henning, 2001).

2.5.1. Milking environmental hygiene

Studies have examined risk factors associated with bacterial contamination of raw bulk milk. Verbeke *et al.* (2014) observed that certain hygiene interventions involving the milking equipment and the milking environment in a Flemish dairy herd were associated with a decrease in bacterial counts including coliforms.

Husbandry and milking procedures do impact bacterial and somatic cell counts in bulk tank milk on dairy farms, but consistent application of a few hygienic practices could significantly improve the microbiological quality of raw milk (Elmoslemany *et al.*, 2010).

> *Adequate milking hygiene is a GHP; however, the degree of support as an intervention specifically for the control of STEC in raw milk was low.*

2.5.2. Udder hygiene

The epithelium of the udder and teats can be contaminated with STEC (Fremaux *et al.*, 2006). Pre-dipping teats into a disinfectant solution followed by drying is a GHP that has been shown to be an effective teat skin sanitation against many potential mastitis and spoilage organisms (Galton, Petrsson and Merrill, 1986; Gibson *et al.*, 2008). Use of an automated teat scrubber (rotating brushes) for chlorine dioxide disinfection and drying also effectively reduced bacterial loads on teats (Baumberger, Guarín and Ruegg, 2016; Elmoslemany *et al.*, 2010). While none of these basic sanitation processes have been examined against STEC specifically, pre-milking treatment should be used in combination with the control of contamination from the environment, milking equipment and water.

> *Udder hygiene is a GHP and the degree of support as an intervention specifically for the control of STEC in raw milk was medium.*

2.5.3. Milk storage temperature and hygiene

Temperature control and hygiene in the farming and processing environments as well as during transportation from the farm are critical factors in the commercial milk supply chain and can significantly affect the microbiological quality of raw milk prior to packaging and sale for drinking or for use in the manufacture of raw milk cheeses.

Depending on the size of the herd and bulk tank capacity, the bulk tank is a key storage area for milk collected from one or more milkings. Bulk tanks are usually made of stainless steel which aids in cleaning to remove build-up of milk films (and potentially biofilms), but the tanks still have areas that are less accessible to cleaning (e.g. valves, outlet ports and gaskets) (EFSA, 2015). The formation and presence of biofilms in milk bulk tanks is a concern, but should also be considered in all steps between the milking equipment and the tankers hauling milk from the farms (Weber *et al.*, 2019).

During processing, temperatures ≥ 6 °C and/or extended storage of raw milk were associated with a significant increase in bacterial counts (ICMSF, 2001 and 2011;

Vithanage *et al.*, 2017). An increased duration of cold storage from long distance transportation or including water rinses of tanks in between tanker loads had no negative effects on raw milk quality. Although not yet at the level of an accepted standard, a full cleaned-in-place approach where tankers are cleared every 24 h with water rinses along with use of a sanitizer treatment between loads, reduced the presence of surface-associated bacteria in the tankers (Darchuk, Meunier-Goddik and Waite-Cusic, 2015).

Temperature change, extended storage time and initial bacterial counts in raw milk during collection, storage and transportation have been associated with increased counts of *E. coli* in raw milk. Cooling, hygiene practices and reduced storage and transportation times can reduce *E. coli* and other indicator organism counts, but no specific evidence was found regarding the effect of these practices on STEC.

> *Cleanliness of the bulk tank and temperature control of milk during storage are GHP; however, the degree of support as an intervention specifically for the control of STEC in raw milk was low.*

2.6. ANIMAL TRANSPORTATION

Pertinent to control of STEC, animal transportation issues that must be addressed include hide contamination (mixing of cattle from different farms, cleanliness of trucks and loading areas), the impact of animal stress from excessive temperature, humidity, loading density and the duration of transport, as all of these can affect the colonization and faecal excretion of STEC.

2.6.1. Feed withdrawal prior to slaughter

Cattle should not be fed for a minimum of 8 h to a maximum of 12 h pre-slaughter to avoid very full gastrointestinal tracts, which are more likely to rupture during the evisceration process and increase the potential for spread of STEC onto carcasses. Feed withdrawal reduces faecal output and contamination of the environment and hides, yet because it also results in decreased volatile fatty acid concentrations and increased pH in the gastrointestinal tract, feed withdrawal can lead to an increase in STEC excretion (Pointon, Kiermeier and Fegan, 2012). Most research on the effect of fasting has been conducted on-farm and no studies have investigated the effect of fasting on the reduction of STEC excretion while the cattle are in lairage pens.

> *The degree of support for feed withdrawal prior to slaugher as an intervention specifically for the control of STEC in raw beef was low.*

2.6.2. Duration of transportation

Bach *et al.* (2004) evaluated the effects of pre-conditioning (attenuating stress effects) and the duration of animal transport from the pasture on faecal excretion of *E. coli* and STEC O157:H7 by beef calves. Authors observed an increase in excretion and suggested that the calves' susceptibility to infection from the environment was likely elevated by the stresses of weaning, transport, and relocation. Faecal excretion increased with lack of pre-conditioning and long periods of transport. Supporting this study, Dewell *et al.* (2008) observed that lots (groups) of cattle that were transported for long distances (>160 km) had twice the risk of having STEC positive hide samples at slaughter compared with cattle transported a shorter distance. Increased STEC excretion with longer distance of transport is also supported by the study by Arthur *et al.* (2007). But, other studies did not find an association between transportation stress or general heat stress and temperature with faecal excretion of STEC O157:H7 in cattle (Brown-Brandl *et al.*, 2009; Pfeiffer *et al.*, 2009). Transport density and trailer design, such as multi-level trailers, may also influence the degree that hides are positive with STEC O157:H7 resulting from faecal-coat contamination (e.g. hide tag or dag, or more colloquially "dingleberry") (Stanford *et al.*, 2011).

> *The degree of support for transportation time as an intervention specifically for the control of STEC in raw beef was low.*

References

Ahmad, A., Nagaraja, T.G. & Zurek, L. 2007. Transmission of *Escherichia coli* O157:H7 to cattle by house flies. *Preventative Veterinary Medicine*, 80: 74–81. doi: 10.1016/j.prevetmed.2007.01.006

Arthur, T.M., Bosilevac, J.M., Brichta-Harhay, D.M., Guerini, M.N., Kalchayanan, N., Shackleford, S.D., Wheeler, T.L. & Koohmaraie, M. 2007. Transportation and lairage environment effects on prevalence, numbers, and diversity of *Escherichia coli* O157:H7 on hides and carcasses of beef cattle at processing. *Journal of Food Protection*, 70: 280–286. doi: 10.4315/0362-028x-70.2.280

Arthur, T.M., Kalchayanand, N., Agga, G.E., Wheeler, T.L. & Koohmaraie, M. 2017. Evaluation of bacteriophage application to cattle in lairage at beef processing plants to reduce *Escherichia coli* O157:H7 prevalence on hides and carcasses. *Foodborne Pathogens and Disease*, 14: 17–22. doi: 10.1089/fpd.2016.2189

Bach, S.J., McAllister, T.A., Mears, G.J. & Schwartzkof-Genswein, K.S. 2004. Long-haul transport and lack of preconditioning increases fecal shedding of *Escherichia coli* and *Escherichia coli* O157:H7 by calves. *Journal of Food Protection*, 67: 672–678. doi: 10.4315/0362-028x-67.4.672

Bach, S.J., Selinger, J., Stanford, K. & McAllister, T.A. 2005a. Effect of supplementing corn- or barley-based feedlot diets with canola oil on faecal shedding of *Escherichia coli* O157:H7 by steers. *Journal of Applied Microbiology*, 98: 464–75. doi: 10.1111/j.1365-2672.2004.02465.x

Bach, S.J., Stanford, K. & McAllister, T.A. 2005b. Survival of *Escherichia coli* O157:H7 in feces from corn- and barley-fed steers. *FEMS Microbiology Letters*, 252: 25–33. doi: 10.1016/j.femsle.2005.08.030

Baumberger, C., Guarín, J.F. & Ruegg, P.L. 2016. Effect of two different premilking teat sanitation routines on reduction of bacterial counts on teat skin of cows on commercial dairy farms. *Journal of Dairy Science*, 99: 2915–2929. doi: 10.3168/jds.2015-10003

Beauvais, W., Gart, E.V., Bean, M., Blanco, A., Wilsey, J., McWhinney, K., Bryan, L., Krath, M., Yang, C.Y., Alvarez, D.M., Paudyal, S., Bryan, K., Stewart, S., Cook, P.W., Lahodny, G., Baumgarten, K, Gautam, R., Nightingale, K., Lawhon, S.D., Pinedo, P. & Ivane, R. 2018. The prevalence of *Escherichia coli* O157:H7 fecal shedding in feedlot pens is affected by the water-to-cattle ratio: A randomized controlled trial. *PLOS ONE*, 13: e0192149. doi: 10.1371/journal.pone.0192149

Berg, J.L., McAllister, T.A., Bach, S.J., Stillborn, R.P., Hancock D.D. & LeJeune, J.T. 2004. *Escherichia coli* O157:H7 excretion by commercial feedlot cattle fed either barley- or corn-based finishing diets. *Journal of Food Protection*, 67: 666–671. doi: 10.4315/0362-028x-67.4.666

Berry, E.D., Wells, J.E., Varel, V.H., Hales K.E. & Kalchayanand, N. 2017. Persistence of *Escherichia coli* O157:H7 and total *Escherichia coli* in feces and feedlot surface manure from cattle fed diets with and without corn or sorghum wet distillers grains with solubles. *Journal of Food Protection*, 80: 1317–1327. doi: 10.4315/0362-028X. JFP-17-018

Besser, T.E., Schmidt, C.E., Shah, D.H. & Shringi, S. 2014 "Preharvest" food safety for *Escherichia coli* O157 and other pathogenic Shiga toxin-producing strains. *Microbiology Spectrum*, 2: 21–28. doi: 10.1128/microbiolspec.EHEC-0021-2013

Besser, T.E., Schmidt, C.E., Shah, D.H. & Shringi, S. 2015. "Preharvest" food safety for *Escherichia coli* O157 and other pathogenic Shiga toxin-producing strains. In *Enterohemorrhagic Escherichia coli and Other Shiga toxin-Producing E. coli. American Society of Microbiology*, 2(5): 437–456. doi: 10.1128/microbiospec. EHEC-0021-2013

Bibbal, D., Loukiadis, E., Kérourédan, M., Ferré, F., Dilasser, F., Peytavin de Garam, C., Cartier, P., Oswald, E., Gay, E., Auvray, F. & Brugère, H. 2015. Prevalence of carriers of Shiga toxin-producing *Escherichia coli* serotypes O157:H7, O26:H11, O103:H2, O111:H8 and O145:H28 in French slaughtered adult cattle. *Applied and Environmental Microbiology*, 81: 1397–1405. doi: 10.1128/AEM.03315-14

Blaustein, R.A., Pachepsky, Y.A., Shelton, D.R. & Hill, R.L. 2015. Release and removal of microorganisms from land-deposited animal waste and animal manures: a review of data and models. *Journal of Environmental Quality*, 44: 1338–1354. doi: 10.2134/jeq2015.02.0077

Bolton, D.J., O'Neill, C.J. & Fanning, S. 2011. A preliminary study of *Salmonella*, Verocytotoxigenic *Escherichia coli/Escherichia coli* O157 and *Campylobacter* on four mixed farms. *Zoonoses and Public Health*, 59(3): 217–228. doi: 10.1111/j.1863-2378.2011.01438.x

Bosilevac, J.M., Shackelford, S.D., Brichta, D.M. & Koohmaraie, M. 2005. Efficacy of ozonated and electrolyzed oxidative waters to decontaminate hides of cattle before slaughter. *Journal of Food Protection*, 68: 1393–1398. doi: 10.4315/0362-028x-68.7.1393

Bosilevac, J.M., Wang, R., Luedtke, B.E., Hinkley, S., Wheeler, T.L. & Koohmaraie, M. 2017. Characterization of enterohemorrhagic *Escherichia coli* on veal hides and carcasses. *Journal of Food Protection*, 80: 136–145. doi: 10.4315/0362-028X. JFP-16-247

Braden, K.W., Blanton Jr, J.R., Allen, V.G., Pond, K.R. & Miller, M.F. 2004. Ascophyllum nodosum supplementation: A preharvest intervention for reducing *Escherichia coli* O157:H7 and *Salmonella* spp. in feedlot steers. *Journal of Food Protection*, 67: 1824–1828. doi: 10.4315/0362-028x-67.9.1824

Branham, L.A., Carr, M.A., Scott, C.B. & Callaway, T.R. 2005. *E. coli* O157 and *Salmonella* spp. in white-tailed deer and livestock. *Current Issues in Intestinal Microbiology*, 6: 25–29. PMID: 16107036

Brown-Brandl, T.M., Berry, E.D., Wells, J.E., Arthur, T.M. & Nienaber, J.A. 2009. Impacts of individual animal response to heat and handling stresses on *Escherichia coli* and *E. coli* O157:H7 fecal shedding by feedlot cattle. *Foodborne Pathogens and Disease*, 6: 855–864. doi: 10.1089/fpd.2008.0222

Callaway, T.R., Anderson, R.C., Genovese, K.J., Poole, T.L., Anderson, T.J., Byrd, J.A., Kubena, L.F. & Nisbet, D.J. 2002. Sodium chlorate supplementation reduces *E. coli* O157:H7 populations in cattle. *Journal of Animal Science*, 80(6): 1683-1689. doi: 10.2527/2002.8061683x

Callaway, T.R., Stahl, C.H., Edrington, T.S., Genovese, K.J., Lincoln, L.M., Anderson, R.C., Lonergan, S.M., Poole, T.L., Harvey, R.B. & Nisbet, D.J. 2004. Colicin concentrations inhibit growth of *Escherichia coli* O157:H7 in vitro. *Journal of Food Protection*, 67: 2603–2607. doi: 10.4315/0362-028x-67.11.2603

Callaway, T.R., Carr, M.A., Edrington, T.S., Anderson, R.C. & Nisbet, D.J. 2009. Diet, *Escherichia coli* O157:H7, and cattle: A review after 10 years. *Current Issues in Molecular Biology*, 11: 67–80. PMID: 19351974

Callaway, T.R., Carroll, J.A., Arthington, J.D., Edrington, T.S., Anderson, R.C., Rossman, M.L., Carr, M.A., Genovese, K.J., Ricke, S.C., Crandall, P. & Nisbet, D.J. 2011. *Escherichia coli* O157:H7 populations in ruminants can be reduced by orange peel product feeding. *Journal of Food Protection*, 74: 1917–1921. doi: 10.4315/0362-028X.JFP-11-234

Callaway, T.R., Edrington, T.S., Loneragan, G.H., Carr, M.A. & Nisbet, D.J. 2013. Shiga toxin-producing *Escherichia coli* (STEC) ecology in cattle and management based options for reducing fecal shedding. *Agriculture Food and Analytical Bacteriology*, 3: 39–69

Castro, V.S., Carvalho, R.C.T., Conte-Junior, C.A. & Figuiredo, E.E.S. 2017. Shiga-toxin roducing *Escherichia coli*: pathogenicity, supershedding, diagnostic methods, occurrence, and foodborne outbreaks. *Comprehensive Reviews in Food Science and Food Safety*, 16: 1269-1280. doi: 10.1111/1541-4337.12302

Cernicchiaro, N., Pearl, D.L., McEwen, S.A., Harpster, L., Homan, H.J., Linz, G.M. & LeJeune, J.T. 2012. Association of wild bird density and farm management factors with the prevalence of *E. coli* O157 in dairy herds in Ohio (2007-2009). *Zoonoses and Public Health*, 59: 320–329. doi: 10.1111/j.1863-2378.2012.01457.x

Cízek, A., Alexa, P., Literák, I., Hamrík, J., Novák, P. & Smola, J. 1999. Shiga toxin-producing *Escherichia coli* O157 in feedlot cattle and Norwegian rats from a large-scale farm. *Letters in Applied Microbiology*, 6: 435–9. doi: 10.1046/j.1365-2672.1999.00549.x

Cobbold, R. & Desmarchelier, P. 2002. Horizontal transmission of Shiga toxin-producing *Escherichia coli* within groups of dairy calves. *Applied and Environmental Microbiology*, 68(8): 4148–4152. doi: 10.1128/AEM.68.8.4148-4152.2002

Darchuk, E.M., Meunier-Goddik, L. & Waite-Cusic, J. 2015. Microbial quality of raw milk following commercial long-distance hauling. *Journal of Dairy Science*, 98: 8572–6. doi: 10.3168/jds.2015-9771

Dawson, D.E., Keung, J.H., Napoles, M.G., Vella, M.R., Chen, S., Sanderson, M.W. & Lanzas, C. 2018. Investigating behavioral drivers of seasonal Shiga-Toxigenic *Escherichia coli* (STEC) patterns in grazing cattle using an agent-based model. *PLOS ONE*, 13(10): e0205418. doi: 10.1371/journal.pone.0205418

Depenbusch, B.E., Nagaraja, T.G., Sargeant, J.M., Drouillard, J.S., Loe, E.R. & Corrigan, M.E. 2008. Influence of processed grains on fecal pH, starch concentration, and shedding of *Escherichia coli* O157 in feedlot cattle. *Journal of Animal Science*, 86: 632–639. doi: 10.2527/jas.2007-0057

Dewell, G.A., Simpson, C.A., Dewell, R.D., Hyatt, D.R., Belk, K.E., Scanga, J.A., Morley, P.S., Grandin, T., Smith, G.C., Dargatz, D.A., Wagner, B.A. & Salman, M.D. 2008. Impact of transportation and lairage on hide contamination with *Escherichia coli* O157 in finished beef cattle. *Journal of Food Protection*, 71: 1114–1118. doi: 10.4315/0362-028x-71.6.1114

Diez-Gonzalez, F., Callaway, T.R., Kizoulis, M.G. & Russell, J.B. 1998. Grain feeding and the dissemination of acid-resistant *Escherichia coli* from cattle. *Science*, 281: 1666–1668. doi: 10.1126/science.281.5383.1666

Duffy, G. 2003. Verocytoxigenic *Escherichia coli* in animal faeces, manures and slurries. *Journal of Applied Microbiology*, 94 Suppl: 94S–103S. doi: 10.1046/j.1365-2672.94.s1.11.x

Edrington, T. S., Callaway, T. R., Bischoff, K. M., Genovese, K. J., Elder, R. O., Anderson, R. C. & Nisbet, D. J. 2003. Effect of feeding the ionophores monensin and laidlomycin propionate and the antimicrobial bambermycin to sheep experimentally infected with *E. coli* O157:H7 and *Salmonella* Typhimurium. *Journal of Animal Science*, 81: 553–560. doi: 10.2527/2003.812553x

Edrington, T.S., Schultz, C.L., Genovese, K.J., Callaway, T.R, Looper, M.L., Bischoff, K.M., McReynolds, J.L., Anderson, R.C. & Nisbet, D.J. 2004. Examination of heat stress and stage of lactation (early versus late) on fecal shedding of *E. coli* O157:H7 and *Salmonella* in dairy cattle. *Foodborne Pathogens and Disease*, 1: 114–119. doi: 10.1089/153531404323143639

Edrington, T.S., Carter, B.H., Friend, T.H., Hagevoort, G.R., Poole, T.L., Callaway, T.R., Anderson, R.C. & Nisbet, D.J. 2009a. Influence of sprinklers, used to alleviate heat stress, on faecal shedding of *E. coli* O157:H7 and *Salmonella* and antimicrobial susceptibility of *Salmonella* and Enterococcus in lactating dairy cattle. *Letters in Applied Microbiology*, 48: 738–743. doi: 10.1111/j.1472-765X.2009.02603.x

Edrington, T.S., Farrow, R.L., Loneragan, G.H., Ives, S.E., Engler, M.J., Wagner, J.J., Corbin, M.J., Platter, W.J., Yates, D., Hutcheson, J.P., Zinn, R.A, Callaway, T.R., Anderson R.C. & Nisbet, D.J. 2009b. Influence of beta-agonists (Ractopamine HCl and Zilpaterol HCl) on fecal shedding of *Escherichia coli* O157:H7 in feedlot cattle. *Journal of Food Protection*, 72: 2587–2591. doi: 10.4315/0362-028X-72.12.2587

EFSA (European Food Safety Authority). 2015. Scientific opinion of the panel of Biological Hazards on the public health risks related to the consumption of raw drinking milk. *EFSA Journal*, 13(1): 3940. doi: 10.2903/j.efsa.2015.3940

Ekong, P.S., Sanderson, M.W. & Cernicchiaro, N. 2015. Prevalence and concentration of *Escherichia coli* O157 in different seasons and cattle types processed in North America: A systematic review and meta-analysis of published research. *Preventative Veterinary Medicine*, 121: 74–85. doi: 10.1016/j.prevetmed.2015.06.019

Elder, R.O., Keen, J.E., Wittum, T.E., Callaway, T.R., Edrington, T.S., Anderson, R.C. & Nisbet, D.J. 2002. Intervention to reduce fecal shedding of enterohemorrhagic *Escherichia coli* O157:H7 in naturally infected cattle using neomycin sulfate. *Journal of Animal Science*, 80 (Suppl. 1): 15 (Abstr.). ISSN: 0021-8812

Ellis-Iversen, J., Smith, R.P., Van Winden, S., Paiba, G.S., Watson, E., Snow, L.C. & Cook, A.J. 2008. Farm practices to control *E. coli* O157 in young cattle-a randomised controlled trial. *Veterinary Research*, 39: 3–15. doi: 10.1051/vetres:2007041

Elmoslemany, A.M., Keefe, G.P., Dohoo, I.R., Wichtel, J.J., Stryhn, H. & Dingwell, R.T. 2010. The association between bulk tank milk analysis for raw milk quality and on-farm management practices. *Preventative Veterinary Medicine*, 95: 32–40. doi: 10.1016/j.prevetmed.2010.03.007

Fingermann, M., Avila, L., Belén De Marco, M., Vázquez, L., Di Biase, D.N., Müller, A.V., Lescano, M., Dokmetjian, J.C., Castillo, S.F & Pérez Quiñoy, J.L. 2018. OMV-based vaccine formulations against Shiga toxin producing *Escherichia coli* strains are both protective in mice and immunogenic in calves. *Human Vaccines and Immunotherapeutics*, 14(9): 2208–2213. doi: 10.1080/21645515.2018.1490381

Fox, J.T., Depenbusch, B.E., Drouillard, J.S. & Nagaraja, T.G. 2007. Dry-rolled or steam-flaked grain-based diets and fecal shedding of *Escherichia coli* O157 in feedlot cattle. *Journal of Animal Science*, 85: 1207–1212. doi: 10.2527/jas.2006-079

Frank, C., Kapfhammer, S., Werber, D., Stark, K. & Held, L. 2008. Cattle density and Shiga toxin-producing *Escherichia coli* infection in Germany: Increased risk for most but not all serogroups. *Vector-borne Zoonotic Disease*, 8: 635–643. doi: 10.1089/vbz.2007.0237

Fremaux, B., Raynaud, S., Beutin, L. & Rozand, C.V. 2006. Dissemination and persistence of Shiga toxin-producing *Escherichia coli* (STEC) strains on French dairy farms. *Veterinary Microbiology*, 117: 180-91. doi: 10.1016/j.vetmic.2006.04.030

French, E., Rodriguez-Palacios, A. & Lejeune, J.T. 2010. Enteric bacterial pathogens with zoonotic potential isolated from farm-raised deer. *Foodborne Pathogens and Disease*, 7: 1031–1037. doi: 10.1089/fpd.2009.0486

Galton, D.M., Petersson, L.G. & Merrill, W.G. 1986. Effects of pre-milking udder preparation practices on bacterial counts in milk and on teats. *Journal of Dairy Science*, 69: 260–266. doi: 10.3168/jds.S0022-0302(86)80396-4

Garber, L., Wells, S., Schroeder-Tucker, L. & Ferris, K. 1999. Factors associated with fecal shedding of verotoxin-producing *Escherichia coli* O157 on dairy farms. *Journal of Food Protection*, 62: 307–12. doi: 10.4315/0362-028x-62.4.307

Gautam, R., Bani-Yaghoub, M., Neill, W.H., Döpfer, D., Kaspar, C. & Ivanek, R. 2011. Modeling the effect of seasonal variation in ambient temperature on the transmission dynamics of a pathogen with a free-living stage: example of *Escherichia coli* O157:H7 in a dairy herd. *Preventative Veterinary Medicine*, 102: 10–21. doi: 10.1016/j.prevetmed.2011.06.008

Gibson, H., Sinclair, L.A., Brizuela, C.M., Worton, H.L. & Protheroe, R.G. 2008. Effectiveness of selected premilking teat-cleaning regimes in reducing teat microbial load on commercial dairy farms. *Letters in Applied Microbiology*, 46: 295–300. doi: 10.1111/j.1472-765X.2007.02308.x

Gunn, G.J., McKendrick, J., Ternent, H.E., Thomson-Carterd, F., Foster, G. & Synge, B.A. 2007. An investigation of factors associated with the prevalence of verocytotoxin producing *Escherichia coli* O157 shedding in Scottish beef cattle. *The Veterinary Journal*, 174: 554–564. doi: 10.1016/j.tvjl.2007.08.024

Hales, K.E., Wells, J.E., Berry, E.D., Kalchayanand, N., Bono, J.L. & Kim, M. 2017. The effects of monensin in diets fed to finishing beef steers and heifers on growth performance and fecal shedding of *Escherichia coli* O157:H7. *Journal of Animal Science*, 95: 3738–3744. doi: 10.2527/jas.2017.1528

Hancock, D.D., Besser, T.E., Rice, D.H., Ebel, E.D., Herriott, D.E. & Carpenter, L.V. 1998. Multiple sources of *Escherichia coli* O157 in feedlots and dairy farms in the Northwestern USA. *Preventative Veterinary Medicine*, 35: 11–19. doi: 10.1016/s0167-5877(98)00050-6

Hancock, D. D., Besser, T. E., Gill, C. & Bohach, C. H. 1999. Cattle, hay, and *E. coli*. *Science*. 284:49-50. doi: 10.1126/science.284.5411.49g

Hovde, C. J., Austin, P. R., Cloud, K. A., Williams, C. J. & Hunt, C. W. 1999. Effect of cattle diet on *Escherichia coli* O157:H7 acid resistance. *Applied and Environmental Microbiology*, 65:3233-3235. doi: 10.1128/aem.65.7.3233-3235.1999

ICMSF (International Commission on Microbiological Specifications for Foods). 2001. Microorganisms in Foods 7: Microbiological testing in food safety management. NY: Kluwer Academic / Plen Publishers.

ICMSF. 2011. Microorganisms in Foods 8: Use of data for assessing process control and product acceptance. NY: Springer.

Jacob, M.E., Fox, J.T., Narayanan, S.K., Drouillard, J.S., Renter, D.G. & Nagaraja, T.G. 2008. Effects of feeding wet corn distiller's grains with solubles with or without monensin and tylosin on the prevalence and antimicrobial susceptibilities of fecal food-borne pathogenic and commensal bacteria in feedlot cattle. *Journal of Animal Science*, 86: 1182–1190. doi: 10.2527/jas.2007-0091

Jacob, M.E., Callaway, T.R. & Nagaraja, T.G. 2009. Dietary interactions and interventions affecting *Escherichia coli* O157 colonization and shedding in cattle. *Foodborne Pathogens and Disease*, 6: 785–792. doi: 10.1089/fpd.2009.0306

Jayarao, B.M. & Henning, D.R. 2001. Prevalence of foodborne pathogens in bulk tank milk. *Journal of Dairy Science*, 84: 2157–2162. doi: 10.3168/jds.S0022-0302(01)74661-9

Jeon, S.J., Elzo, M., DiLorenzo, N., Lamb, G.C. & Jeong, K.C. 2013. Evaluation of animal genetic and physiological factors that affect the prevalence of *Escherichia coli* O157 in cattle. *PLOS ONE*, 8: e55728. doi: 10.1371/journal.pone.0055728

Keen, J. E., Uhlich, G. A., & Elder, R. O. 1999. Effects of hay-and grain-based diets on fecal shedding in naturally-acquired enterohemorrhagic *E. coli* (EHEC) O157 in beef feedlot cattle. In *Procs. 80th Conf. Research Workers in Animal Diseases*. Chicago, IL. (Abstr.).

Kimmitt, P.T., Harwood, C.R. & Barer, M.R. 2000. Toxin gene expression by Shiga toxin-producing *Escherichia coli*: the role of antibiotics and the bacterial SOS response. *Emerging Infectious Diseases*, 6: 458–465. doi: 10.3201/eid0605.000503

Köhler, B., Karch, H. & Schmidt, H. 2000. Antibacterials that are used as growth promoters in animal husbandry can affect the release of Shiga-toxin-2-converting bacteriophages and Shiga toxin 2 from *Escherichia coli* strains. *Microbiology*, 146: 1085–1090. doi: 10.1099/00221287-146-5-1085

Larzábal, M., Cataldi, A.A. & Vilte, D.A. 2019. Human and veterinary vaccines against pathogenic *Escherichia coli*. In: M. S. Erjavec, ed. *The Universe of Escherichia coli*. IntechOpen. doi: 10.5772/intechopen.73717

LeJeune, J.T., Besser, T.E. & Hancock, D.D. 2001. Cattle water troughs as reservoirs of *Escherichia coli* O157. *Applied Environmental Microbiology*, 67: 3053–57. doi: 10.1128/AEM.67.7.3053–3057.2001

LeJeune, J.T., Besser, T.E., Rice, D.H., Berg, J.L., Stilborn, R.P. & Hancock, D.D. 2004. Longitudinal study of fecal shedding of *Escherichia coli* O157:H7 in feedlot cattle: predominance and persistence of specific clonal types despite massive cattle population turnover. *Applied and Environmental Microbiology*, 70: 377–84. doi: 10.1128/aem.70.1.377-384

Liu, H., Niu, Y.D., Meng, R., Wang, J., Li, J., Johnson, R.P., McAllister, T.A. & Stanford, K. 2015. Control of *Escherichia coli* O157 on beef at 37, 22 and 4° C by T5-, T1-, T4-and O1-like bacteriophages. *Food Microbiology*, 51: 69–73. doi: 10.1016/j. fm.2015.05.001

Mather, A.E., Innocent, G.T., McEwen, S.A., Reilly, W.J., Taylor, D.J., Steele, W.B., Gunn, G.J., Ternent, H.E., Reid, S.W.J. & Mellor, D.J. 2007. Risk factors for hide contamination of Scottish cattle at slaughter with *Escherichia coli* O157. *Preventive Veterinary Medicine*, 80: 257–270. doi: 10.1128/AEM.00770-08

McAllister, T.A., Bach, S.J., Stanford, K. & Callaway, T.R. 2006. Shedding of *Escherichia coli* O157:H7 by cattle fed diets containing monensin or tylosin. *Journal of Food Protection*, 69: 2075–2083. doi: 10.4315/0362-028x-69.9.2075

Money, P., Kelly, A.F., Gould, S.W., Denholm-Price, J., Threlfall, E.J. & Fielder, M.D. 2010. Cattle, weather and water: mapping *Escherichia coli* O157:H7 infections in humans in England and Scotland. *Environmental Microbiology*, 12: 2633–44. doi: 10.1111/j.1462-2920.2010.02293.x

Morrow, J.L., Mitloehner, F.M., Johnson, A.K., Galyean, M.L., Dailey, J.W., Edrington, T.S., Anderson, R.C., Genovese, K.J., Poole, T.L., Duke, S.E. & Callaway, T.R. 2005. Effect of water sprinkling on incidence of zoonotic pathogens in feedlot cattle. *Journal of Animal Science*, 83: 1959–1966. doi: 10.2527/2005.8381959x

Munns, K.D., Selinger, L., Stanford, K., Selinger, L.B. & McAllister, T.A. 2014. Are super-shedder feedlot cattle really super? *Foodborne Pathogen and Disease*, 11: 329–331. doi: 10.1089/fpd.2013.1621

Munns, K.D., Zaheer, R., Xu, Y., Stanford, K., Laing, C.R., Gannon, V.P.J., Selinger, L.B. & McAllister, T.A. 2015. Comparative genomics of *Escherichia coli* O157:H7 isolated from super-shedder and low-shedder cattle. *PLOS ONE*, 11: e0151673. doi: 10.1371/ journal.pone.0151673

Ongeng, D., Geeraerd, A.H., Springael, D., Ryckeboer, J., Muyanja, C. & Mauriello, G. 2015. Fate of *Escherichia coli* O157:H7 and *Salmonella* enterica in the manure-amended soil-plant ecosystem of fresh vegetable crops: a review. *Critical Reviews in Microbiology*, 41(3): 273–294. doi: 10.3109/1040841X.2013.829415

Paddock, Z.D., Walker, C.E., Drouillard, J.S. & Nagaraja, T.G. 2011. Dietary monensin level, supplemental urea, and ractopamine on fecal shedding of *Escherichia coli* O157:H7 in feedlot cattle. *Journal of Animal Science*, 89: 2829–2835. doi: 10.2527/jas.2010-3793

Pointon, A., Kiermeier, A. & Fegan, N. 2012. Review of the impact of pre-slaughter feed curfews of cattle, sheep and goats on food safety and carcase hygiene in Australia. *Food Control*, 26: 313–321. doi: 10.1016/j.foodcont.2012.01.034

Riley, D.G., Gray, J.T., Loneragan, G.H., Barling, K.S. & Chase Jr., C.C. 2003. *Escherichia coli* O157:H7 prevalence in fecal samples of cattle from a southeastern beef cow-calf herd. *Journal of Food Protection*, 66: 1778–1782. doi: 10.4315/0362-028x-66.10.1778

Rice, D.H., Sheng, H.Q., Wynia, S.A. & Hovde, C.J. 2003. Rectoanal mucosal swab culture is more sensitive than fecal culture and distinguishes *Escherichia coli* O157:H7-colonized cattle and those transiently shedding the same organism. *Journal of Clinical Microbiology*, 41(11): 4924–4929. doi: 10.1128/JCM.41.11.4924–4929.2003

Sabouri, S., Sepehrizadeh, Z., Amirpour-Rostami, S. & Skurnik, M. 2017. A minireview on the in vitro and in vivo experiments with anti-*Escherichia coli* O157:H7 phages as potential biocontrol and phage therapy agents. *International Journal of Food Microbiology*, 243: 52–57. doi: 10.1016/j.ijfoodmicro.2016.12.004

Sánchez, S., Martínez, R., García, A., Vidal, D., Blanco, J., Blanco, M., Blanco, J.E., Mora, A., Herrera León, S., Echeita, A., Alonso, J.M. & Rey, J. 2010. Detection and characterisation of O157:H7 and non-O157 Shiga toxin-producing *Escherichia coli* in wild boars. *Veterinary Microbiology*, 143: 420–423. doi: 10.1016/j.vetmic.2009.11.016

Sanderson, M.W., Sargeant, J.M., Shi, X., Nagaraja, T.G., Zurek, L. & Alam, M.J. 2006. Longitudinal emergence and distribution of *Escherichia coli* O157 genotypes in a beef feedlot. *Applied and Environmental Microbiology*, 72: 7614–7619. doi: 10.1128/AEM.01412-06

Schuehle-Pfeiffer, C.E., King, D.A., Lucia, L.M., Cabrera Diaz, E., Acuff, G.R., Randel, R.D., Welsh, T.H., Oliphint, C. & Vann, S. 2009. Influence of transportation stress and animal temperament on fecal shedding of *Escherichia coli* O157:H7 in feedlot cattle. *Meat Science*, 81: 300–306. doi: 10.1016/j.meatsci.2008.08.005

Schulz, S., Stephan, A., Hahn, S., Bortesi, L., Jarczowski, F., Bettmann, U., Paschke, A.K., Tuse, D., Stahl, C.H., Giritch, A. & Gleba, Y. 2015. Broad and efficient control of major foodborne pathogenic strains of *Escherichia coli* by mixtures of plant-produced colicins. *Proceedings of the National Academy of Sciences (USA)*, 112: E5454–60. doi: 10.1073/pnas.1513311112

Smith, D., Blackford, M., Younts, S., Moxley, R., Gray, J., Hungerford, L., Milton, T. & Klopfenstein, T. 2001. Ecological relationships between the prevalence of cattle shedding *Escherichia coli* O157:H7 and characteristics of the cattle or conditions of the feedlot pen. *Journal of Food Protection*, 64: 1899–903. doi: 10.4315/0362-028X-64.12.1899

Smith, D.R., Klopfenstein, T., Moxley, R.A., Milton, C.T., Hungerford, L.L. & Gray, J.T. 2002. An evaluation of three methods to clean feedlot water tanks. *The Bovine Practitioner*, 36: 1–4.

Smith, D.R. 2015. Vaccination of cattle against *Escherichia coli* O157: H7. In: V. Sperandio, C. Hovde Bohach, eds. Enterohemorrhagic *Escherichia coli* and Other Shiga Toxin-Producing *E. coli*, pp. 505 -529. *American Society of Microbiology*. ISBN-10: 1555818781.

Snedeker, K.G., Campbell, M. & Sargeant, J.M. 2012. A systematic review of vaccinations to reduce the shedding of *Escherichia coli* O157 in the faeces of domestic ruminants. *Zoonoses and Public Health*, 59: 126–138. doi: 10.1111/j.1863-2378.2011.01426.x

Stacey, K.F., Parsons, D.J., Christiansen, K.H. & Burton, C.H. 2007. Assessing the effect of interventions on the risk of cattle and sheep carrying *Escherichia coli* O157:H7 to the abattoir using a stochastic model. *Preventative Veterinary Medicine*, 79: 32–45. doi: 10.1016/j.prevetmed.2006.11.007

Stanford, K., Bryan, M., Peters, J., Gonzalez, L.A., Stephens, T.P. & Schwartzkopf-Genswein, K.S. 2011. Effects of long- or short-haul transportation of slaughter heifers and cattle liner microclimate on hide contamination with *Escherichia coli* O157. *Journal of Food Protection*, 74: 1605–1610. doi: 10.4315/0362-028X.JFP-11-154

Stenkamp-Strahm, C., Lombard, J.E., Magnuson R.J., Linke, L.M., Magzamen, S., Urie, N.J., Shivley, C.B. & McConnel, C.S. 2018. Preweaned heifer management on US dairy operations: Part IV. Factors associated with the presence of *Escherichia coli* O157 in preweaned dairy heifers. *Journal of Dairy Science*, 101: 9214–9228. doi: 10.3168/jds.2018-14659

Strachan, N.J.C., Fenlon, D.R. & Ogden, I.D. 2001. Modelling the vector pathway and infection of humans in an environmental outbreak of *Escherichia coli* O157:H7. *FEMS Microbiology Letters*, 203: 69–73. doi: 10.1111/j.1574-6968.2001.tb10822.x

Synge, B.A., Chase-Topping, M.E., Hopkins, G.F., McKendrick, I.J., Thomson-Carter, F., Gray, D., Rusbridge, S.M., Munro, F.I., Foster, G. & Gunn, G.J. 2003. Factors influencing the shedding of verocytotoxin-producing *Escherichia coli* O157 by beef suckler cows. *Epidemiology and Infection*, 130(2): 301–312. doi: 10.1017/S0950268802008208

Talley, J.L., Wayadande, A.C., Wasala, L.P., Gerry, A.C., Fletcher, J., DeSilva, U. & Gilliland, S.E. 2009. Association of *Escherichia coli* O157:H7 with filth flies (Muscidae and Calliphoridae) captured in leafy greens fields and experimental transmission of *E. coli* O157:H7 to spinach leaves by house flies (diptera: Muscidae). *Journal of Food Protection*, 72: 1547–1552. doi: 10.4315/0362-028x-72.7.1547

Van Kessel, J.S., Nedoluha, P.C., Williams-Campbell, A., Baldwin, R.L. & McLeod, K.R. 2002. Effects of ruminal and postruminal infusion of starch hydrolysate or glucose on the microbial ecology of the gastrointestinal tract in growing steers. *Journal of Animal Science*, 80: 3027–3034.

Varela, N.P., Dick, P. & Wilson, J. 2012. Assessing the existing information on the efficacy of bovine vaccination against *Escherichia coli* O157: H7–A systematic review and meta-analysis. *Zoonoses and Public Health*, 60(4): 253–268. doi: 10.1111/j.1863-2378.2012.01523.x

Venegas-Vargas, C., Henderson, S., Khare, A., Mosci, R.E., Lehnert, J.D., Singh, P., Ouellette, L.M., Norby, B., Funk, J.A., Rust, S., Bartlett, P.C., Grooms, D. & Manning, S.D. 2016. Factors associated with Shiga toxin-producing *Escherichia coli* shedding by dairy and beef cattle. *Applied and Environmental Microbiology*, 82: 5049–56. doi: 10.1128/AEM.00829-16

Verbeke, J., Piepers, S., Supré, K. & De Vliegher, S. 2014. Pathogen-specific incidence rate of clinical mastitis in Flemish dairy herds, severity, and association with herd hygiene. *Journal of Dairy Science*, 97: 6926–6934. doi:10.3168/jds.2014-8173

Vidovic, S. & Korber, D.R. 2006. Prevalence of *Escherichia coli* O157 in saskatchewan cattle: Characterization of isolates by using random amplified polymorphic DNA PCR, antibiotic resistance profiles, and pathogenicity determinants. *Applied and Environmental Microbiology*, 72: 4347–4355. doi: 10.1128/AEM.02791-05

Vithanage, N.R., Dissanayake M., Bolze, G., Palombo, E.A., Yeager, T.R. & Datt, N. 2017. Microbiological quality of raw milk attributable to prolonged refrigeration conditions. *Journal of Dairy Research*, 84: 92–101. doi: 10.1017/S0022029916000728

Wang, L., Qu, K., Li, X., Cao, Z., Wang, X., Li, Z., Song, Y. & Xu, Y. 2017. Use of bacteriophages to control *Escherichia coli* O157:H7 in domestic ruminants, meat products, and fruits and vegetables. *Foodborne Pathogens and Disease*, 14: 483–493. doi: 10.1089/fpd.2016.2266

Wang, O., Liang, G., McAllister, T.A., Plastow, G., Stanford, K., Selinger, B. & Guan, L.L. 2016. Comparative transcriptomic analysis of rectal tissue from beef steers revealed reduced host immunity in *Escherichia coli* O157:H7 super-shedders. *PLOS ONE*, 11: e0151284. doi: 10.1371/journal.pone.0151284

Weber, M., Liedtke, J., Plattes, S. & Lipski, A. 2019. Bacterial community composition of biofilms in milking machines of two dairy farms assessed by a combination of culture-dependent and -independent methods. *PLOS ONE*, 14: e0222238. doi: 10.1371/journal.pone.0222238

Wells, J.E., Shackelford, S.D., Berry, E.D., Kalchayanand, N., Bosilevac, J.M. & Wheeler, T.L. 2011. Impact of reducing the level of wet distillers grains fed to cattle prior to harvest on prevalence and levels of *Escherichia coli* O157:H7 in feces and on hides. *Journal of Food Protection*, 74: 1611–1617. doi: 10.4315/0362-028X.JFP-11-160

Wells, J.E., Berry, E.D., Kim, M., Shackelford, S.D. & Hales, K.E. 2017. Evaluation of commercial-agonists, dietary protein, and shade on fecal shedding of *Escherichia coli* O157:H7 from feedlot cattle. *Foodborne Pathogens and Disease*, 14: 649–655. doi: 10.1089/fpd.2017.2313

Wetzel, A.N. & LeJeune, J.T. 2006. Clonal dissemination of *Escherichia coli* O157:H7 subtypes among dairy farms in Northeast Ohio. *Applied and Environmental Microbiology*, 72: 2621–2626. doi: 10.1128/AEM.72.4.2621-2626

Wisener, L.V., Sargeant, J.M., O'Connor, A.M., Faires, M.C. & Glass-Kaastra, S.K. 2015. The use of direct-fed microbials to reduce shedding of *Escherichia coli* O157 in beef cattle: A Systematic Review and Meta-analysis. *Zoonoses and Public Health*, 62: 75–89. doi: 10.1111/zph.12112

Wolin, M.J. 1969. Volatile fatty acids and the inhibition of *Escherichia coli* growth by rumen fluid. *Applied Microbiology*, 17: 83–87. doi: 10.1128/am.17.1.83-87.1969

Xu, Y., Dugat-Bony, E., Zaheer, R., Selinger, L., Barbieri, R., Munns, K., McAllister, T.A. & Selinger, L.B. 2014. *Escherichia coli* O157:H7 super-shedder and non-shedder feedlot steers harbour distinct fecal bacterial communities. *PLOS ONE*, 9: e98115. doi: 10.1371/journal.pone.0098115

Zaheer, R., Dugat-Bony, E., Holman, D., Cousteix, E., Xu, Y., Munns, K., Selinger, L.J., Barbieri, R., Alexander, T., McAllister, T.A. & Selinger, L.B. 2017. Changes in bacterial community composition of *Escherichia coli* O157:H7 super-shedder cattle occur in the lower intestine. *PLOS ONE*, 12: e0170050. doi: 10.1371/journal.pone.0170050

Zhao, T., Tkalcic, S., Doyle, M.P., Harmon B.G., Brown, C.A. & Zhao, P.J. 2003. Pathogenicity of enterohemorrhagic *Escherichia coli* in neonatal calves and evaluation of fecal shedding by treatment with probiotic *Escherichia coli*. *Journal of Food Protection*, 66: 924–30. doi: 10.4315/0362-028x-66.6

Zhou, M., Hünerberg, M., Chen, Y., Reuter, T., McAllister, T.A., Evans, F., Critchley, A.T. & Guan, L.L. 2018. Air-dried brown seaweed, *Ascophyllum nodosum*, alters the rumen microbiome in a manner that changes rumen fermentation profiles and lowers the prevalence of foodborne pathogens. *mSphere*, 3: e00017-18. doi: 10.1128/mSphere.00017-18

3

Processing control strategies for STEC in beef

This section focuses on interventions applied during meat processing to reduce the prevalence and concentration of STEC on beef carcasses and meat products and prevent further cross-contamination of other meat products. Processing stages considered include animal receiving and lairage, slaughter (hide removal, and carcass evisceration, trimming, and dressing) and carcass pre-chilling and chilling. A summary of processing control measures for STEC in beef and their degree of support (high, medium, low), based on scientific evidence, is available in Annex 2.

3.1. LAIRAGE

It is well recognized that the hide of ruminants presented for slaughter is the most important source of microbial contamination for carcasses and the processing environment (Cernicchiaro *et al.*, 2020). Most of the microorganisms found on the hide are of faecal origin with some originating from the farm environment. Pathogens such as STEC may be present in the faecal material, hence on the hides of the cattle, posing a risk for cross-contamination of other animals in the lairage.

Upon arrival at the processing plant, cattle are unloaded and directed through common alleys to lairage pens to be held until slaughtered. Upon exit from the lairage pens, cattle are directed through more common alleys before reaching the area for stunning and shackling. These common use alleys can spread contamination between the animals. Proper lairage pen and working facility hygiene is a GHP that will not reduce STEC carriage in an animal, but may reduce contact transfer of STEC on the hides of cattle.

3.1.1. Lairage cleanliness

Animals presented for slaughter are a source of microbial contamination for lairage areas and pens (Small et al., 2003). A study by Small et al. (2007) demonstrated that the use of pressure washing with water and quaternary ammonium sanitizers and/ or steam under pressure reduced E. coli and Enterobacteriaceae levels in lairage surfaces, ranging from 0.9-5.8 $\log_{10}$ CFU/cm^2.

If the lairage pens and alleys that have previously housed other lots of animals are not cleaned or sanitized in between animal lots, hides contaminated with STEC during transport can transfer STEC within and between animal lots. The cleanliness of the animal intake, lairage and animal handling environments (washing trailers, cattle handling facilities, holding pens in between use) are important to ensure that hides contain a minimum amount of faecal material to reduce STEC cross-contamination. Similarly, Dewell et al. (2008) reported that cattle lots held in STEC O157:H7-positive lairage pens were eight times more likely to have hides that test positive at slaughter than cattle held in lairage pens that tested negative for STEC O157:H7.

Maintaining hygienic environmental conditions (e.g. cleaning, disinfecting and dry conditions), and limiting the amount of time spent in lairage are common animal health and animal welfare management practices. Implementing these steps can reduce animal stress, hide contamination, faecal output, potential gastrointestinal content spillage, and carcass contamination during processing. The efficacy of interventions (primarily in the form of GHP) applied during animal intake and lairage is unclear and likely dependent on the amount of time the animals spent in lairage, the level of stress experienced by the cattle and the animal density of the lairage pens.

> *Lairage cleanliness is a GHP; however, the degree of support as an intervention specifically for the control of STEC in raw beef was low.*

3.1.2. Livestock cleanliness

Recognizing the role of dirty animals in introducing microbial contamination into the slaughter plant, many countries have introduced policies around cleanliness of livestock (Gagaoua et al., 2022). The management of animals classified as dirty can be used to reduce the risk of microbial contamination. Meat processors can implement a range of interventions and GHP's from holding cattle for a period of time on straw in lairage to changing logistics in slaughter (e.g. slaughtering the dirtiest animals at the end of the day to reduce cross contamination; clipping hide after kill and before hide removal; and slowing the slaughter line speed to allow for more care with hide removal). Limited studies that examined animal dirtiness have generally showed a direct correlation between visual cleanliness of cattle hide and lower microbiological counts for aerobic organisms (aerobic plate counts [APC], *Enterobacteriaceae* and *E. coli*) on derived carcasses (Blagojevic *et al.*, 2012). Studies conducted in commercial processing plants in the United States of America, found a significant association between hide cleanliness scores with STEC O157:H7 prevalence after controlling for season of sampling (Cernicchiaro *et al.*, 2020), and similar associations with hide cleanliness were observed with STEC O145 (Schneider *et al.*, 2018; Antic *et al.*, 2010b; Nastasijevic, Mitrovic and Buncic, 2008; Smith *et al.*, 2005; Van Donkersgoed *et al.*, 1997).

> *Livestock cleanliness is a GHP; however, the degree of support as an intervention specifically for the control of STEC in raw beef was low.*

3.1.3. Holding animals in lairage

To comply with animal welfare regulations, the amount of time that animals are held in lairage should be minimized. Researchers have investigated changes in faecal and hide-on prevalence and concentration of STEC from feedlot to lairage pens and then subsequently, during later processing stages, however, studies specifically evaluating the effect of the time spent in lairage pens on controlling STEC excretion have not been reported. It is recognized however, that spending less time in holding pens can reduce faecal contamination among pen-mates by decreasing exposure to other animals defecating and/or to contaminated facilities. Yet different hide decontamination and disinfection procedures remain useful and a key issue to reduce faecal contamination and keep animals dry and mud-free.

> *The degree of support for miminizing time spent in liarage as an intervention specifically for the control of STEC in raw beef was low.*

3.2. HIDE DECONTAMINATION

Most hide decontamination interventions applied before stunning consist of live animal hide washes. Washing cattle with water, ozonated or electrolyzed water or water with the addition of chemicals can reduce visible hide contamination and the level of generic organisms (Bosilevac *et al.*, 2005); however, most of these studies were laboratory-based and the results were inconsistent in terms of specifically reducing potential spread of STEC from hides to carcass surfaces. Washes and hair removal before or after stunning can reduce visible hide contamination and can lower levels of generic microbes.

3.2.1. Bacteriophage

Bacteriophage cocktails specific for STEC O157:H7, may be applied as a spray or a mist, to the hides of live cattle in the holding pens, up to 4 h before slaughter and hide removal. Such treatments are predominantly used in warm summer months. A laboratory-based study showed a 1.50 $\log_{10}$ CFU reduction in STEC O157:H7 on hides treated with bacteriophage (Coffey *et al.*, 2011); however, when phage treatment was done under commercial beef plant conditions, STEC O157:H7 prevalence on hides was only reduced from 57.6–51.8 percent, indicating no significant impact (Arthur *et al.*, 2017). The efficacy of bacteriophage treatment on hides remains unclear and further in-plant studies are needed. Different regulatory positions in the different countries may also affect implementation. A further limitation of this intervention is that at present, most of the phages developed for use only targets STEC O157:H7.

> *The degree of support for the application of bacteriophage to hides as an intervention specifically for the control of STEC in raw beef was low.*

3.2.2. Hide washes with ambient or hot water, organic acids, and other chemicals

A range of hide washes may be applied, most commonly after stunning but before hide removal. Washing cattle hides using pressure hoses for 3 min removed faecal contamination and decreased STEC O157:H7 prevalence on inoculated hides (Byrne *et al.*, 2000). The use of ozonated and electrolyzed oxidizing water reduced the concentration of generic organisms (Bosilevac *et al.*, 2005). The use of chlorinated water (Arthur *et al.*, 2007; Carlson *et al.*, 2008), sodium hydroxide wash with a chlorinated (1 ppm) water rinse (Bosilevac *et al.*, 2006) or washes with 1.0 percent cetylpyridinium chloride (CPC) (Bosilevac *et al.*, 2004) reduced the prevalence of STEC O157:H7 on hide-on surfaces. However, another study reported an increase in the concentration of generic organisms after the application of water with chemicals (Mies *et al.*, 2004).

Electrolyzed-oxidized (EO) water applied to the hide is subject to inactivation as a result of interaction with organic matter and UV light, as is chlorine (Stevenson *et al.*, 2004). Using inoculated hides, EO water reduced STEC O157:H7 hide concentrations by up to 4.3 $\log_{10}$ CFU/100 cm^2, and reduced hide prevalence from 82 percent to 35 percent. The primary application for EO water currently is within processing and post-processing environments (Bosilevac *et al.*, 2005).

Addition of 220 ppm hypobromous acid to wash water reduced STEC O157:H7 hide prevalence from 25.3 to 10.1 percent (Schmidt *et al.*, 2012). Meta-analysis of the literature found that the use of sodium hydroxide or lactic acid was effective for hide decontamination, but the incorporation of water washes along with the antimicrobial washes were largely ineffective, as water diluted and removed antimicrobials (Zhilyaev *et al.*, 2017).

Other factors to consider include, as antimicrobial washes drain down the carcass, they may redistribute microbial contamination to other parts of the carcass. Acid washes can potentially select for acid-resistant microorganisms that may accelerate spoilage and the appearance of undesirable products. The need for specialized equipment/infrastructure to implement and the possibility for increases in equipment corrosion. Exposure to chemicals also raises environmental and employees' health and safety concerns.

> *The degree of support for hide washes using ambient or hot water, organic acids and other chemicals as an intervention specifically for the control of STEC in raw beef was low.*

3.2.3. Hide clipping, coating and chemical dehairing

Hair removal may be beneficial in reducing overall contamination of carcasses, however, its efficacy at reducing STEC is controversial. Concerns about environmental and worker health and safety limits the utility of hair removal. Coating hides with shellac, which immobilizes bacteria, could also reduce pathogen transmission to the carcass. The use of food-grade resin in ethanol (shellac) to coat inoculated hides was reported to successfully reduce STEC O157:H7 prevalence on hides (Antic *et al.*, 2010a) and when applied under commercial conditions, also reduced the level of generic organisms (Antic, Blagojevic and Buncic, 2011).

Even if proven to be efficacious, the environmental issues associated with waste disposal, lack of large-scale studies, need for specific infrastructure to implement, personnel health concerns associated with exposure to the chemicals used, and the scarce evidence supporting their use, the practicality of these interventions is considered limited.

> *The degree of support for hair removal as an intervention specifically for the control of STEC in raw beef was low.*

3.3. SLAUGHTER AND DRESSING

Pre-evisceration measures to prevent contamination of carcasses with fecal material or remove visible faecal material from carcasses are hygienic practices that in principle, reduce general microbial contaminants, including pathogen prevalence and concentration. Important factors for consideration in their use are worker skill, equipment operational maintenance, potential for re-distribution of the carcass contamination or cross-contamination and destruction or loss of the carcass surface and meat.

3.3.1. Speed of processing

The speed at which animals are moved along the processing line has been reported to have an effect on microbial levels on carcasses following dressing (Sheridan, 1998). But, the evidence is conflicting in studies from different countries and data comparisons are complicated by the variable conditions within and among the plants (e.g. efficiency of management systems, worker skill and working conditions, time allowed) and the use of decontamination systems that could mask the effects of rapid line speeds.

No consistent evidence was found on the impact of processing line speed on the microbial load or STEC contamination on dressed carcasses, but this factor varies with worker and processing practices in an establishment. However, plants may decide to alter line speeds based on the amount of mud coat on the hide to reduce microbial cross contamination.

> *The degree of support for reducing processing speed as an intervention specifically for the control of STEC in raw beef was low.*

3.3.2. Hide removal

Contamination from hide to carcass may occur each time the hide is incised through its surface (Huynh *et al.*, 2016). Contamination can also occur from the operator's hands, utensils and equipment used to create opening cuts and to pull the hide.

Slaughter facilities should rely on good dressing procedures during de-hiding to prevent or minimize any contact with the carcass (Gagaoua *et al.*, 2022). Workers should sanitize the knives in a hot water bath between carcasses to minimize cross-contamination. Studies showed that holding knives and steels in water at 82 °C for at least 30 s or an equivalent combination of conditions, resulted in a 2 $\log_{10}$ CFU reduction of *E. coli* on the knives (Eustace *et al.*, 2007; McEvoy *et al.*, 2001). To minimize contamination more efficiently, the use of a two-knife system is recommended, as one knife can be held in hot water while the other knife is being used (EFSA, 2013).

The use of hide pullers is the final step in hide removal and allows for a clean pull of the hide from the carcass without damaging the hide. Downward hide pullers are the most common type but there are some limited users of upward pullers. As the hide is removed, microorganisms on the hide are released into the air as part of droplets and particulate matter that may settle on the carcasses. Kang *et al.* (2019) reported that a downward pulling system resulted in lower bacterial loads on the carcasses while Kennedy *et al.* (2014) found no significant differences. However, the designs of these two studies were not directly comparable.

Evidence regarding the choice of up- or downward hide pulling system to minimize microbial contamination was evaluated as low. Knife and steel hygiene procedures are considered GHP and widely implemented to minimize transfer of microbial contamination from hides to carcasses and between carcasses.

Good dressing procedures are GHP; however, the degree of support for hide removal practices as an intervention specifically for the control of STEC in raw beef was low.

3.3.3. Pre-evisceration and evisceration processes

Pre-evisceration and evisceration of the beef carcass include the removal of the organs, respiratory tract, rumen and other parts of the gastrointestinal tract. In this process, the abdominal cavity is opened, and the contents are removed by cutting away the fat, membrane, and connective tissue attaching the abdominal contents to the carcass. Good dressing practices are particularly important during evisceration to ensure that the rumen and intestinal tract are not punctured, which could result in gross carcass contamination with digesta. Good dressing procedures should also ensure that each end of the gastrointestinal tract of the animal is sealed off before evisceration, to prevent spillage of gastrointestinal content, to prevent carcass contamination with faecal microorganisms. "Bunging" and "weasanding" are the two practices that are used to seal off the rectum and the esophagus, respectively and ensure that the connective tissue attaching both the esophagus and rectum are separated from the carcass. These two practicesmay be associated with lower STEC contamination of carcasses (Stopforth *et al.*, 2006; Sheridan, 1998). Yet, in a meta-analysis on *E. coli* interventions, Greig *et al.* (2012) found that while these pre-evisceration practices are commonly used, there were few studies that reported on their effectiveness to prevent contamination, and this precluded their inclusion in the meta-analysis of available research evidence.

Practices to prevent the leakage of contents from the gastrointestinal tract prior to evisceration are commonly used and recommended as GHPs. When appropriately applied, they generally reduced contamination of carcass with gastrointestinal microorganisms that may include pathogens such as STEC.

Good dressing procedures are GHP, and the degree of support for evisceration practices that prevent leakage of gastrointestinal tract contents as an intervention specifically for the control of STEC in raw beef was medium.

3.3.4. Removal of visible faecal material from carcass

A variety of practices may be carried out to remove visible faecal material, and the associated faecal microorganisms from the carcass, though most of these are not STEC-specific. Water washing, targeted trimming, and steam vacuuming have all been used to reduce visible faecal material contamination on carcasses, though most of these methodologies generally reduce bacterial levels on the carcass.

Water washing of carcass

As a GHP, pre-evisceration washing of carcasses may be carried out just after hide removal. Cold or warm water, or organic acid washes have been used to remove visible carcass contamination (Antic, 2018). In a meta-analysis by Greig *et al.* (2012) on the interventions applied to beef carcasses to control *E. coli*, hot water (74 °C applied with a nozzle pressure of 700 lb/in^2 for 5.5 s) and a 2 percent lactic acid wash warmed in an online spray cabinet, reduced the prevalence of STEC O157:H7 by 81 percent and 35 percent respectively; but sequential treatment of these interventions had no additional benefit (Bosilevac *et al.*, 2006). Depending on the water volume and pressure used, this intervention may actually redistribute contamination on the carcass surface.

> *Removal of visible faecal material from carcasses is a GHP; however, the degree of support for the use of water washing as an intervention specifically for the control of STEC in raw beef was low.*

Trimming of carcass

After carcass splitting and spinal cord removal, knife trimming may be carried out to remove visible faecal contamination. This is a GHP commonly adopted by commercial beef processing plants and has been shown to reduce APC of spoilage organisms from 3.0–4.3 log$_{10}$ CFU/cm^2 (Castillo *et al.*, 1998a; Prasai *et al.*, 1995). Horchner *et al.* (2020) reported that at commercial plants, trimming reduced total viable bacterial counts (TVC) by 0.44 log$_{10}$ CFU/cm^2 and the prevalence of NTS *E. coli* by 29.1 percent.

In a study with inoculated carcasses, trimming reduced STEC O157:H7 by 3 log$_{10}$ CFU/100 cm^2; however, spread of contamination was observed to occur which required further tissue removal (Castillo *et al.*, 1998a). When trimming was combined with another intervention like hot water, lactic acid, or steam vacuum, TVCs was reduced by 0.61 log$_{10}$ CFU/cm^2 and the prevalence of *E. coli* by 36.8 percent (Castillo *et al.*, 1998b).

Trimming was considered a GHP, but it was noted that there are conflicting results on its impact in reducing TVCs, *E. coli* and inoculated STEC O157:H7.

The efficacy of this intervention is very dependent on the workers' skill level and operational maintenance of the equipment. Trimming may also contribute to possible redistribution of contamination on the carcass or cross-contamination of other carcasses from knives and personnel hands/gloves. Trimming also resulted in losses of carcass meat, including surface fat and tissue that can lead to drying, degradation of meat cuts, and may impact product aesthetics for the consumers, thus it has low evidence for use as an STEC reduction strategy.

> *Removal of visible faecal material from carcasses is a GHP and the degree of support for trimming as an intervention specifically for the control of STEC in raw beef was medium.*

Steam vacuuming

Steam vacuuming utilizes a hand-held device comprised of a vacuum wand with a hot spray nozzle, which delivers water at 82-95 °C to the carcass surface under pressure, while simultaneously vacuuming the area to remove faecal material. This treatment has been reported to reduce STEC O157:H7 by 5 $\log_{10}$ CFU/cm^2 on experimentally inoculated beef (Dorsa, Cutter and Siragusa, 1996). Commercial steam vacuum systems have been reported to reduce *E. coli* by 2.8-5.5 $\log_{10}$ CFU/cm^2 (Castillo *et al.*, 1998b; Moxley and Acuff, 2014). On naturally contaminated carcasses under commercial processing conditions, steam vacuuming after carcass trimming was reported to reduce mean TVCs by 0.4-0.9 $\log_{10}$ CFU/cm^2 (Hochreutener *et al.*, 2017). In a meta-analysis, the average reduction of *E. coli* on beef carcasses was 3.1 $\log_{10}$ CFU/cm^2 (Zhilyaev *et al.*, 2017).

The effectiveness of steam vacuuming depends on worker diligence and skill, operational maintenance of the equipment, exposure time and application temperature. It has been reported that a non-permanent, discolouration of the carcass surface can occur. Based on a meta-analysis of literature and its applicability under commercial conditions (in-plant studies), steam vacuuming remains a valuable tool to reduce surface faecal contamination from carcasses.

> *Removal of visible faecal material from carcasses is a GHP, and the degree of support for the use steam vacuuming as an intervention specifically for the control of STEC in raw beef was medium to high.*

3.3.5. Rinsing of head and cheek meat

Head and cheek meats are excised in a separate step from the carcass dressing. Head and cheek meat have been found to have high levels of microbial contamination that occur either naturally, due to contaminants being washed down on inverted and vertically railed carcass during washing, or due to poor

GHP during processing and chilling. Washing animal heads using water or water treated with chemicals has been proposed as a treatment to reduce STEC O157:H7 contamination of the associated head meat. There is some evidence of STEC O157:H7 reduction specifically on masseter muscles (cheek meat) following head washing (Kalchayanand *et al.*, 2008). Most of these results were obtained from challenge studies or simulations and their practicality requires further evaluation with studies in a slaughter plant.

> *Rinsing of head and cheek meat is a GHP; however, the degree of support for the use of this practice as an intervention specifically for the control of STEC in raw beef was low.*

3.4. PRE-CHILLING

Interventions may be applied to the carcass post-dressing and pre-chill to reduce microbial contamination on the carcass surface (Gagaoua *et al.*, 2022). These interventions may be physical, chemical or biological.

Based on a meta-analysis of literature, there is good (high quality of evidence) evidence that the use of hot potable water carcass wash, steam pasteurization and 24 h air chilling and combination of these, are effective in reducing NTS *E. coli* and potentially pathogen contaminants on beef.

3.4.1. Hot water wash

Water at varying temperatures may be used to wash the dressed carcass at the pre-chill stage. The temperature achieved on the carcass surface will have the most impact in terms of microbial reductions and it is affected by the temperature, the pressure and the volume of water applied, and the distance between the spray nozzles and the carcass (Gagaoua *et al.*, 2022). The application of hot water to a carcass may result in a bleached discolouration of the meat, but this generally disappears after chilling. The use of very high-pressure water may also drive bacteria into the carcass tissue rather than removing them.

A commercial hot water wash cabinet set at 74 °C for 5.5 s reduced TVC and *Enterobacteriaceae* counts by 2.7 $\log_{10}$ CFU/cm^2 and reduced the prevalence of carcasses that were positive for STEC O157:H7 by 81 percent (Bosilevac *et al.*, 2006). A meta-analysis of interventions for *E. coli* on beef carcasses (Zhilyaev *et al.*, 2017) estimated that the efficacy of wash water against *E. coli* was increased by 0.014 $\log_{10}$ CFU/cm^2 per °C increase in temperature. Using a cocktail of STEC serogroups inoculated onto beef flank, hot water (85 °C) in a spray cabinet at 1.05 kg/cm^2 (60 cycles) pressure reduced STEC levels by 3.3 and 4.2 $\log_{10}$ CFU/cm^2

(Kalchayanand *et al.*, 2012). Similarly, *stx* gene prevalence was reduced on carcass following hot water spray (82 °C) (Signorini *et al.*, 2018).

The degree of support for the use of hot water carcass wash at pre-chill as an intervention specifically for the control of STEC in raw beef was high.

3.4.2. Steam pasteurization

Steam (100 °C) has a higher heat capacity than water at the same temperature and therefore, should better penetrate the carcass meat surface to target microorganisms present. At the carcass surface, a temperature of $\geq$ 82.2 °C for 6 s to 11 s is reached using steam. Steam may result in carcass discolouration, but acceptable colour is restored after 24 h of chilling. Steam pasteurization has been demonstrated to effect significant reductions in the concentration of TVC and coliforms (below detectable levels) on pre-chill carcasses. A commercial trial showed a reduction in E coli levels (0.5 $\log_{10}$ CFU/cm^2) at rump sites only, along with reductions in *Enterobacteriaceae* (0.8 $\log_{10}$ CFU/cm^2) levels at all carcass sites examined (Minihan *et al.*, 2003). When used on inoculated pre-rigor beef, STEC O157:H7 was reduced by 3.5 $\log_{10}$ CFU/cm^2 (Phebus *et al.*, 1997).

The degree of support for the use of hot steam pasteurization at pre-chill as an intervention specifically for the control of STEC in raw beef was high.

3.4.3. Organic acids

Organic acids (e.g. lactic, formic, propionic, citric, fumaric, L-ascorbic, acetic and mixtures) may be applied to carcasses after trimming and inspection but before chilling. Internationally, solutions of lactic or acetic acids (1 to 3 percent) are commonly used chemical interventions in commercial plants for beef dressing (Gagaoua *et al.*, 2022) and can effect reductions of 0.02 to 3 $\log_{10}$ CFU/cm^2 for APC, *Enterobacteriaceae*, coliforms and NTS *E. coli* (Dormedy *et al.*, 2000; Bosilevac *et al.*, 2006; Signorini, 2018).

Lactic acid produces reductions of 2–3 $\log_{10}$ CFU/cm^2 of STEC O157:H7 on beef carcasses (Ransom *et al.*, 2003). Reductions in *E. coli* levels from treatment with 2 percent solutions of lactic, acetic and citric acids applied manually or automatically, ranged from 0.08–0.83 $\log_{10}$ CFU/cm^2 depending on acid type, temperature and the mode of application. In most cases, automatic application had greater impact than manual. Organic acids have been shown to be most effective when applied as a warm rinse (50 °C to 55 °C) (Acuff, 2005). A 3 percent solution of lactic acid at 55 °C and applied automatically, gave a 1.03 $\log_{10}$ CFU/cm^2 reduction in microbial levels and also a significant reduction (29.3 percent) in *stx* gene prevalence on carcasses (Signorini *et al.*, 2018).

There is a lot of variability in the literature in terms of the cited reductions in STEC that can be achieved through the use of organic acids. This is mainly due to differences in the concentrations and types of acids used by different researchers, the method of application, the types of samples tested, and the initial microbial load of samples.

Carcass surfaces treated with organic acids often display some discolouration of tissue or fat surfaces (Meat Industry Services, 2006). However, discolouration becomes less evident after chilling and may be less apparent if preceded by a hot water (~90 °C) carcass wash.

> *The degree of support for the use of organic acids on carcasses at pre-chill as an intervention specifically for the control of STEC in raw beef was low.*

3.4.4. Oxidizer type antimicrobials

Other agents which act as oxidative biocides can be used as a carcass wash or spray.

Ozone is a water-soluble gas and a strong oxidizing agent which must be generated at the point of use. Application of 0.5 percent ozonated water on beef tissue reduced total bacterial counts by 2.5 $\log_{10}$ CFU/cm^2 (Gorman *et al.*, 1995, 2007). However, a study of the effectiveness of an ozone treatment in reducing STEC O157:H7 and *Salmonella* Typhimurium contamination on hot carcass surfaces show ozone treatment had no significant improvement over a water wash in reducing pathogens on beef carcass surfaces (Castillo *et al.*, 2003). Potential exposure of ozone to workers also poses safety concerns.

Electrolysed water is generated by passing electric current through a dilute saline solution. At present, there is limited evidence of its efficacy to reduce microorganisms on beef carcass.

Peracetic/peroxyaectic acid can be applied to carcasses generally at levels of around 200 ppm. Reported reductions of STEC O157:H7 levels on meat carcasses have varied from 0.7 $\log_{10}$ CFU/cm^2 (King *et al.*, 2005); 1 to 1.4 $\log_{10}$ CFU/cm^2 (Ransom *et al.*, 2003) to 2.2 $\log_{10}$ CFU/cm^2 (Penney *et al.*, 2007). Peracetic and peroxyacetic acids treatments are commonly used in beef processing either alone or in combination with other agents.

Acidified sodium chlorite (ASC) can be used at concentrations between 500 and 1200 ppm as a wash/spray on beef carcasses. There are conflicting reports on its effectiveness ranging from limited reduction in APC and *E. coli* levels (Gill and Badoni, 2004) to 0.6–2.3 $\log_{10}$ CFU/cm^2 reduction of different STEC serogroups (Kalchayanand *et al.*, 2012).

The degree of support for the use of ozonated and electolyzed water and other chemicals on carcasses at pre-chill as an intervention specifically for the control of STEC in raw beef was low.

3.5. CARCASS CHILLING

The biochemical processes and structural changes that occur in beef during the first 24 h post-mortem are critical in determining product quality and palatability (Reid *et al.*, 2017).

The objective of chilling carcasses is to cool the meat quickly enough to prevent bacterial growth, but not so quickly to cause cold shortening (toughening) of the meat. The higher the carcass surface temperatures, greater the likelihood of bacterial growth, including spoilage bacteria (Gagaoua *et al.*, 2022), which may result in higher bacterial counts and shorter product shelf-life. Chilling, obtained by setting a critical limit of ≤ 4 °C surface temperature within 24 h, is considered a critical control point in the Hazard Analysis and Critical Control Point (HACCP) plans of many processing plants.

Conventional air chilling has been reported to reduce levels of APC and indicator microorganisms by 0.5 to 2 $\log_{10}$ CFU/cm^2 on carcasses, but there is some evidence to suggest that this reduction may be an artefact, as the bacteria are only stressed and given appropriate conditions, the stressed bacterial cells may recover (Mellefont, Kocharunchitt and Ross, 2015). Spray chilling of carcasses uses water micro droplets with a chilling regime for approximately 14 h with intermittent spraying cycles. Spray is commonly used for beef and is designed to reduce carcass weight loss. There is no evidence that spray chilling has a substantial effect on microbial populations, including STEC.

The use of spray chilling in combination with oxidizer type of antimicrobials has also been reported. In a laboratory study, aqueous ozone applied as a spray chill on inoculated beef showed significant reductions in STEC O157:H7 and APC levels as compared to spray chilling with water alone (Kalchayanand, Worlie and Wheeler, 2019). Simulated spray chilling with chlorine dioxide (> 20 ppm) and peroxyacetic acid (> 200 ppm) on inoculated beef striploins reduced *E. coli* by ≥ 4 $\log_{10}$ CFU/cm^2 (Kocharunchitt *et al.*, 2020). The use of such combination approaches show potential, but more evidence is needed on their performance under commercial processing conditions.

Spray chilling and incorporating antimicrobial oxidizers is a GHP; however, the degree of support for the use of this practice specifically for the control of STEC in raw beef was low.

References

Acuff, G. R. 2005. Chemical decontamination strategies for meat. In: J.N. Sofos, ed. *Improving the Safety of Fresh Meat*, pp. 351-363. New York, USA, Woodhead Publishing Limited. CRC Press.

Antic, D., Blagojevic, B. & Buncic, S. 2011. Treatment of cattle hides with Shellac solution to reduce hide-to-beef microbial transfer. *Meat Science*, 88: 498–502. doi: 10.1016/j.meatsci.2011.01.034

Antic, D., Blagojevic, B., Ducic, M., Mitrovic, R., Nastasijevic, I. & Buncic, S. 2010a. Treatment of cattle hides with Shellac-in-ethanol solution to reduce bacterial transferability - a preliminary study. *Meat Science*, 85: 77–81. doi: 10.1016/j.meatsci.2009.12.007

Antic, D., Blagojevic, B., Ducic, M., Nastasijevic, I., Mitrovic, R. & Buncic, S. 2010b. Distribution of microflora on cattle hides and its transmission to meat via direct contact. *Food Control*, 21: 1025–1029. doi: 10.1016/j.foodcont.2009.12.022

Antic, D. 2018. A critical literature review to assess the significance of intervention methods to reduce the microbiological load on beef through primary production. Report of FSA Project FS301044. www.food.gov.uk/print/pdf/node/4506

Arthur, T. M., Bosilevac, J. M., Brichta-Harhay, D. M., Kalchayanand, N., Shackelford, S. D., Wheeler, T. L. & Koohmaraie, M. 2007. Effects of a minimal hide wash cabinet on the levels and prevalence of *Escherichia coli* O157:H7 and *Salmonella* on the hides of beef cattle at slaughter. *Journal of Food Protection*, 70: 1076–1079. doi: 10.4315/0362-028X-70.5.1076

Arthur, T. M., Kalchayanand, N., Agga, G. E., Wheeler, T. L. & Koohmaraie, M. 2017. Evaluation of bacteriophage application to cattle in lairage at beef processing plants to reduce *Escherichia coli* O157:H7 prevalence on hides and carcasses. *Foodborne Pathogens and Disease*, 14: 17–22. doi: 10.1089/fpd.2016.2189

Blagojevic, B., Antic, D., Ducic, M. & Buncic, S. 2012. Visual cleanliness scores of cattle at slaughter and microbial loads on the hides and the carcases. *Veterinary Record*, 170: 563. doi: 10.1136/vr.100477

Bosilevac, J. M., Nou, X., Barkocy-Gallagher, G. A., Arthur, T. M. & Koohmaraie, M. 2006. Treatments using hot water instead of lactic acid reduce levels of aerobic bacteria and Enterobacteriaceae and reduce the prevalence of *Escherichia coli* O157:H7 on preevisceration beef carcasses. *Journal of Food Protection*, 69: 1808–1813. doi: 10.4315/0362-028x-69.8.1808

Bosilevac, J. M., Shackelford, S. D., Brichta, D. M. & Koohmaraie, M. 2005. Efficacy of ozonated and electrolyzed oxidative waters to decontaminate hides of cattle before slaughter. *Journal of Food Protection*, 68: 1393–1398. doi: 10.4315/0362-028x-68.7.1393

Bosilevac, J. M., Arthur, T. M., Wheeler, T. L., Shackelford, S. D., Rossman, M., Reagan, J. O. & Koohmaraie, M. 2004. Prevalence of *Escherichia coli* O157 and levels of aerobic bacteria and *Enterobacteriaceae* are reduced when hides are washed and treated with cetylpyridinium chloride at a commercial beef processing plant. *Journal of Food Protection*, 67(4): 646–650. doi: 10.4315/0362-028X-67.4.646

Byrne, C. M., Bolton, D. J., Sheridan, J. J., McDowell, D. A. & Blair, I. S. 2000. The effects of preslaughter washing on the reduction of *Escherichia coli* O157:H7 transfer from cattle hides to carcasses during slaughter. *Letters in Applied Microbiology*, 30(2): 142–145. doi: 10.1046/j.1472-765x.2000.00689.x

Carlson, B. A., Ruby, J., Smith, G. C., Sofos, J. N., Bellinger, G. R., Warren-Serna, W., Centrella, B., Bowling, R. A. & Belk, K. E. 2008. Comparison of antimicrobial efficacy of multiple beef hide decontamination strategies to reduce levels of *Escherichia coli* O157:H7 and *Salmonella*. *Journal of Food Protection*, 71: 2223–2227. doi: 10.4315/0362-028x-71.11.2223

Castillo, A., Dickson, J. S., Clayton, R. P., Lucia, L. M. & Acuff, G. R. 1998a. Chemical dehairing of bovine skin to reduce pathogenic bacteria and bacteria of fecal origin. *Journal of Food Protection*, 61: 623–625. doi: 10.4315/0362-028x-61.5.623

Castillo, A., Lucia, L.M., Goodson, K.J., Savell, J.W. & Acuff, G.R. 1998b. Comparison of water wash, trimming, and combined hot water and lactic acid treatments for reducing bacteria of fecal origin on beef carcasses. *Journal of Food Protection*, 61: 823–828. doi: 10.4315/0362-028x-61.7.823

Castillo, A., McKenzie, K.S., Lucia, L.M. & Acuff, G.R. 2003. Ozone treatment for reduction of *Escherichia coli* O157:H7 and *Salmonella* serotype Typhimurium on beef carcass surfaces. *Journal of Food Protection*, 66: 775–779. doi: 10.4315/0362-028x-66.5.775

Cernicchiaro, N., Oliveira, A.R.S., Hoehn, A., Noll, L.W., Shridhar, P.B., Nagaraja, T.G., Ives, S.E., Renter, D.G. & Sanderson, M.W. 2020. Associations between season, processing plant, and hide cleanliness scores with prevalence and concentration of major Shiga toxin-producing *Escherichia coli* on beef cattle hides. *Foodborne Pathogens and Disease*, 17: 611–619. doi: 10.1089/fpd.2019.2778

Coffey, B., Rivas, L., Duffy, G., Coffey, A., Ross, R.P. & McAuliffe, O. 2011. Assessment of *Escherichia coli* O157:H7-specific bacteriophages e11/2 and e4/1c in model broth and hide environments. *International Journal of Food Microbiology*, 147: 188–194. doi: 10.1016/j.ijfoodmicro.2011.04.001

Dewell, G.A., Simpson, C.A., Dewell, R.D., Hyatt, D.R., Belk, K.E., Scanga, J.A., Morley, P.S., Grandin, T., Smith, G.C., Dargatz, D.A., Wagner, B.A. & Salman, M.D. 2008. Impact of transportation and lairage on hide contamination with *Escherichia coli* O157 in finished beef cattle. *Journal of Food Protection*, 71: 1114–1118. doi:10.4315/0362-028X-71.6.1114

Dormedy, E.S., Brashears, M.M., Cutter, C.N. & Burson, D.E. 2000. Validation of acid washes as critical control points in hazard analysis and critical control point systems. *Journal of Food Protection*, 63: 1676–1680. doi: 10.4315/0362-028x-63.12.1676

Dorsa, W.J., Cutter, C.N. & Siragusa, G.R. 1996. Effectiveness of a steam-vacuum sanitizer for reducing *Escherichia coli* O157:H7 inoculated to beef carcass surface tissue. *Letters in Applied Microbiology*, 23: 61–63. doi: 10.1111/j.1472-765x.1996.tb00029.x

Eustace, I., Midgley, J., Giarrusso, C., Laurent, C., Jenson, I. & Sumner, J. 2007. An alternative process for cleaning knives used on meat slaughter floors. *International Journal of Food Microbiology*, 113: 23–27. doi: 10.1016/j.ijfoodmicro.2006.06.034

EFSA (European Food Safety Authority). 2013. Scientific Opinion of the panel of Biological Hazards on the public health hazards to be covered by inspection of meat (bovine animals). *EFSA Journal*, 11(6): 3266. doi:10.2903/j.efsa.2013.326

Gagaoua, M., Duffy, G., Alvarez, C., Burgess, C.M., Hamill, R., Crofton, E., Botinestean, C., Ferragina, A., Cafferky, J., Mullen, A.M. & Troy, D. 2022. Current research and emerging tools to improve fresh red meat quality. *Irish Journal of Agricultural and Food Research*. doi: 10.15212/ijafr-2020-0141

Gill, C.O. & Badoni, M. 2004. Effects of peroxyacetic acid, acidified sodium chlorite or lactic acid solutions on the microflora of chilled beef carcasses. *International Journal of Food Microbiology*, 91: 43–50. doi: 10.1016/s0168-1605(03)00329-5

Gorman, B.M., Morgan, J.B., Sofos, J.N. & Smith, G.C. 1995. Microbiological and visual effects of trimming and/or spray washing for removal of fecal material from beef. *Journal of Food Protection*, 58: 984–989. doi: 10.4315/0362-028X-58.9.984

Gorman, B.M., Kochevar, S.L., Sofos, J., Morgan, B., Schmidt, G.R. & Smith, G.C. 2007. Changes on beef adipose tissue following decontamination with chemical solutions or water of 35C or 74C. *Journal of Muscle Foods*, 8: 185–197. doi: 10.1111/j.1745-4573.1997.tb00627.x

Greig, J.D., Waddell, L., Wilhelm, B., Wilkins, W., Bucher, O., Parker, S. & Rajić, A. 2012. The efficacy of interventions applied during primary processing on contamination of beef carcasses with *Escherichia coli*: A systematic review-meta-analysis of the published research. *Food Control*, 27: 385–397. doi: 10.1016/j.foodcont.2012.03.019

Hochreutener, M., Zweifel, C., Corti, S. & Stephan, R. 2017. Effect of a commercial steam-vacuuming treatment implemented after slaughtering for the decontamination of cattle carcasses. *Italian Journal of Food Safety*, 6: 6864. doi: 10.4081/ijfs.2017.6864

Horchner, P., Huynh, L., Sumner, J., Vanderlinde, P.B. & Jenson, I. 2020. Performance metrics for slaughter and dressing hygiene at Australian beef export establishments. *Journal of Food Protection*, 83: 996–1001. doi: 10.4315/jfp-19-591

Huynh, L., Jenson, I., Kaur, M., Kiermeier, A., Kocharunchitt, C., Miles, D., Ross, T., Sumner, J. & Vanderlinde, P. 2016. Shelf life of Australian red meat, Meat & Livestock Australia, Australia, pp. 1-182. ISBN 9781740362399. www.mla.com.au/globalassets/mla-corporate/research-and-development/program-areas/food-safety/brochures/shelf-life-of-australian-red-meat-2nd-edition.pdf

Kalchayanand, N., Arthur, T.M., Bosilevac, J.M., Brichta-Harhay, D.M., Guerini, M.N., Wheeler, T.L. & Koohmaraie, M. 2008. Evaluation of various antimicrobial interventions for the reduction of *Escherichia coli* O157:H7 on bovine heads during processing. *Journal of Food Protection*, 71: 621–624. doi: 10.4315/0362-028x-71.3.621

Kalchayanand, N., Arthur, T.M., Bosilevac, J.M., Schmidt, J.W., Wang, R., Shackelford, S.D. & Wheeler, T.L. 2012. Evaluation of commonly used antimicrobial interventions for fresh beef inoculated with Shiga toxin-producing *Escherichia coli* serotypes O26, O45, O103, O111, O121, O145, and O157:H7. *Journal of Food Protection*, 75: 1207–1212. doi: 10.4315/0362-028x.jfp-11-53

Kalchayanand, N., Worlie, D. & Wheeler, T. 2019. A novel aqueous ozone treatment as a spray chill intervention against *Escherichia coli* O157:H7 on surfaces of fresh beef. *Journal of Food Protection*, 82: 1874–1878. doi: 10.4315/0362-028x.jfp-19-093

Kang, S., Ravensdale, J., Coorey, R., Dykes, G.A. & Barlow, R. 2019. A comparison of 16S rRNA profiles through slaughter in Australian export beef abattoirs. *Frontiers in Microbiology*, 10: 2747. doi: 10.3389/fmicb.2019.02747

Kennedy, T.G., Giotis, E.S. & McKevitt, A.I. 2014. Microbial assessment of an upward and downward dehiding technique in a commercial beef processing plant. *Meat Science*, 97: 486–489. doi: 10.1016/j.meatsci.2014.03.009

King, D.A., Lucia, L.M., Castillo, A., Acuff, G.R., Harris, K.B. & Savell, W. 2005. Evaluation of peroxyacetic acid as a post-chilling intervention for control of *Escherichia coli* O157:H7 and *Salmonella* Typhimurium on beef carcass surfaces. *Meat Science*, 69: 401–407. doi: 10.1016/j.meatsci.2004.08.010

Kocharunchitt, C., Mellefont, L., Bowman, J.P. & Ross, T. 2020. Application of chlorine dioxide and peroxyacetic acid during spray chilling as a potential antimicrobial intervention for beef carcasses. *Food Microbiology*, 87: 103355. doi: 10.1016/j.fm.2019.103355

McEvoy, J.M., Doherty, A.M., Finnerty, M., Sheridan, J.J., McGuire, L., Blair, I.S., McDowell, D.A. & Harrington, D. 2001. The relationship between hide cleanliness and bacterial numbers on beef carcasses at a commercial abattoir. *Letters in Applied Microbiology*, 30: 390–395. doi: 10.1046/j.1472-765x.2000.00739.x

Meat Industry Services. 2006. Organic Acids. https://meatupdate.csiro.au/new/Organic%20Acids.pdf

Mellefont, L.A., Kocharunchitt, C. & Ross, T. 2015. Combined effect of chilling and desiccation on survival of *Escherichia coli* suggests a transient loss of culturability. *International Journal of Food Microbiology*, 208: 1–10. doi: 10.1016/j.ijfoodmicro.2015.04.024.

Mies, P.D., Covington, B.R., Harris, K.B., Lucia, L.M., Acuff, G.R. & Savell, J.W. 2004. Decontamination of cattle hides prior to slaughter using washes with and without antimicrobial agents. *Journal of Food Protection*, 67: 579–582. doi: 10.4315/0362-028x-67.3.579

Minihan, D., O'Mahony, M., Whyte, P. & Collins, J.D. 2003. An investigation on the effect of transport and lairage on the faecal shedding prevalence of *Escherichia coli* O157 in cattle. *Journal of Veterinary Medicine*, 50: 378–382. doi: 10.1046/j.1439-0450.2003.00674.x

Moxley, R.A. & Acuff, G.R. 2014. Peri- and postharvest factors in the control of Shiga toxin-producing *Escherichia coli* in beef. *Microbiology Spectrum*, 2(6). doi: 10.1128/microbiolspec.EHEC-0017-2013

Nastasijevic, I., Mitrovic, R. & Buncic, S. 2008. Occurrence of *Escherichia coli* O157 on hides of slaughtered cattle. *Letters in Applied Microbiology*, 46: 126–131. doi: 10.1111/j.1472-765X.2007.02270.x

Penney, N., Bigwood, T., Barea, H., Pulford, D., LeRoux, G., Cook, R., Jarvis, G. & Brightwell, G. 2007. Efficacy of a peroxyacetic acid formulation as an antimicrobial intervention to reduce levels of inoculated *Escherichia coli* O157:H7 on external carcass surfaces of hot-boned beef and veal. *Journal of Food Protection*, 70: 200–203. doi: 10.4315/0362-028x-70.1.200

Phebus, R.K., Nutsch, A.L., Schafer, D.E., Wilson, R.C., Riemann, M.J., Leising, J.D., Kastner, C.L., Wolf, J.R. & Prasai, R.K. 1997. Comparison of steam pasteurization and other methods for reduction of pathogens on surfaces of freshly slaughtered beef. *Journal of Food Protection*, 60: 476–484. doi: 10.4315/0362-028x-60.5.476

Prasai, R.K., Phebus, R.K., Garcia Zepeda, C.M., Kastner, C.L., Boyle, A.E. & Fung, D.Y.C. 1995. Effectiveness of trimming and/or washing on microbiological quality of beef carcasses. *Journal of Food Protection*, 58: 1114–1117. doi: 10.4315/0362-028x-58.10.1114

Ransom, J., Belk, K., Sofos, J., Stopforth, J., Scanga, J. & Smith, G. 2003. Comparison of intervention technologies for reducing *Escherichia coli* O157:H7 on beef cuts and trimmings. *Food Protection Trends*, 23: 24–34. ISSN : 1541-9576.

Reid, R., Fanning, S., Whyte, P., Kerry, J. & Bolton, D. 2017. Comparison of hot versus cold boning of beef carcasses on bacterial growth and the risk of blown pack spoilage. *Meat Science*, 125: 46–52. doi: 10.1016/j.meatsci.2016.11.012

Schmidt, J.W., Wang, R., Kalchayanand, N., Wheeler, T.L. & Koohmaraie, M. 2012. Efficacy of hypobromous acid as a hide-on carcass antimicrobial intervention. *Journal of Food Protection*, 75(5): 955–958. doi: 10.4315/0362-028X.JFP-11-433

Schneider, L.G., Stromberg, Z.R., Lewis, G.L., Moxley, R.A. & Smith, R. 2018. Cross-sectional study to estimate the prevalence of enterohaemorrhagic *Escherichia coli* on hides of market beef cows at harvest. *Zoonoses Public Health*, 65: 625–636. doi: 10.1111/zph.12468

Sheridan, J. 1998. Sources of contamination during slaughter and measures for control. *Journal of Food Safety*, 18(4): 321–339. doi: 10.1111/j.1745-4565.1998.tb00223.x

Signorini, M., Costa, M., Teitelbaum, D., Restovich, V., Brasesco, H., García, D., Superno, V., Petroli, S., Bruzzone, M., Arduini, V., Vanzini, M., Sucari, A., Suberbie, G., Maricel, T., Rodríguez, R. & Leotta, G.A. 2018. Evaluation of decontamination efficacy of commonly used antimicrobial interventions for beef carcasses against Shiga toxin-producing *Escherichia coli*. *Meat Science*, 142: 44–51. doi: 10.1016/j.meatsci.2018.04.009

Small, A., James, C., Purnell, G., Losito, P., James, S. & Buncic, S. 2007. An evaluation of simple cleaning methods that may be used in red meat abattoir lairages. *Meat Science*, 75: 220–228. doi: 10.1016/j.meatsci.2006.07.007

Small, A., Reid, C.A. & Buncic, S. 2003. Conditions in lairages at abattoirs for ruminants in southwest England and in vitro survival of *Escherichia coli* O157, *Salmonella* Kedougou, and *Campylobacter jejuni* on lairage-related substrates. *Journal of Food Protection*, 66: 1570–1575. doi: 10.4315/0362-028x-66.9.1570

Smith, D.R., Moxley, R.A., Clowser, S.L., Folmer, J.D., Hinkley, S., Erickson, G.E. & Klopfenstein, T.J. 2005. Use of rope devices to describe and explain the feedlot ecology of *Escherichia coli* O157:H7 by time and place. *Foodborne Pathogens and Disease*, 2: 50–60. doi: 10.1089/fpd.2005.2.50

Stevenson, S.M.L., Cook, S.R., Bach, S.J. & McAllister, T.A. 2004. Effects of storage, water source, bacterial and fecal loads on the efficacy of electrolyzed oxidizing (EO) water for the control of *Escherichia coli* O157:H7. *Journal of Food Protection*, 67: 1377–1383. doi: 10.4315/0362-028x-67.7.1377

Stopforth, J.D., Lopes, M., Shultz, J.E., Miksch, R.R. & Samadpour, M. 2006. Location of bung bagging during beef slaughter influences the potential for spreading pathogen contamination on beef carcasses. *Journal of Food Protection*, 69: 1452–1455. doi: 10.4315/0362-028x-69.6.1452

Van Donkersgoed, J., Jericho, K.W.F., Grogan, H. & Thorlakson, B. 1997. Preslaughter hide status of cattle and the microbiology of carcasses. *Journal of Food Protection*, 60: 1502–1508. doi: 10.4315/0362-028x-60.12.1502

Zhilyaev, S., Cadavez, V., Gonzales-Barron, U., Phetxumphou, K. & Gallagher, D. 2017. Meta-analysis on the effect of interventions used in catt le processing plants to reduce *Escherichia coli* contamination. *Food Research International*, 93: 16–25. doi: 10.1016/j.foodres.2017.01.005

Post-processing control strategies for STEC in beef

The intended use of raw beef is an important factor to consider in the selection and implementation of methods for STEC control. If the product is not intended to remain intact, STEC present on the exterior of meat may be internalized during the non-intact production process, such as grinding and mechanical tenderization. In such cases, cooking to a rare or medium-rare internal temperature may not be sufficient to destroy STEC throughout the product. It is critical therefore, that primal, sub-primal and other cuts intended to be non-intact products should be treated by interventions to reduce or eliminate STEC.

During carcass fabrication, the carcass is broken down into consumer portions, which includes additional product preparation and handling. As all these steps increase surface area of the product, the likelihood of contamination spread is great, therefore the application of inventions to reduce STEC at fabrication can be impactful.

During mechanical tenderization of meats, the needles or blades used in the process of tenderization can physically transfer foodborne pathogens from the surface into the interior of the beef cuts. This has prompted the development of interventions that can reduce internalization of surface STEC (Currie *et al.*, 2019). Some nations have required registered plants to affix a label (Mechanically Tenderized Beef [MTB]) to products and to include safe cooking instructions for the consumers, stating "Cook to a minimum internal temperature of 63 °C" (Health Canada, 2014).

Raw ground beef and ground beef-based products (e.g. hamburger patties), pose a higher risk to human health than intact beef because of its greater contact surface and the higher degree of handling and processing involved with production.

During the mincing/grinding process, microbial transfer from the external surfaces into the mass of the ground beef is likely to occur; therefore, it is important to implement GHP, GMP and HACCP principles as well as intervention measures throughout the ground beef production chain to minimize STEC exposure and contamination. In several nations, all beef used in grinding is required to be tested for contamination by specific STEC serotypes (USDA, 2016, 2017).

Despite all the control measures applied at the previous stages of production, contamination of STEC in ground beef can still be detected, albeit mostly at low concentration. This remains a critical issue however, because of the low infectious dose of STEC, hence interventions still need to be applied at all stages of ground beef production, product manufacturing, packaging and distribution.

Since ground beef is perishable, it is important to apply control measures properly during transport and storage of the carcasses/beef cuts before grinding. Maintaining temperature (< 7 °C) is an important parameter that should be controlled throughout the ground beef production chain to reduce growth of STEC through distribution, retail sale, and until the product reaches the consumer (Duffy *et al.*, 2005). Packaging processes, including interventions, for ground/minced products are also critical for ensuring STEC control. Product labels should contain sufficient information about interventions applied, while also guiding the purchaser with safe handling and preparation guidelines (e.g. use-by dates and the need for thorough cooking on the label).

Although the implementation of the interventions in the post-processing phase are mostly to improve microbial safety of fresh ground beef, other essential parameters must also be considered, such as extension of product shelf-life and consumer acceptance (e.g. maintenance of sensory qualities without altering organoleptic characteristics; inclusion of package labelling regarding the treatment, guidance for safe handling).

The antimicrobial interventions implemented throughout the beef production chain can vary depending on the country's regulation and the volume of production as well as destination of the product (e.g. local consumption vs export market). Intervention strategies used in post-processing should be safe and suitable to be broadly approved by the regulations of different nations.

A summary of post-processing control measures, and combinations of these, for STEC in beef and their degree of support (high, medium, low), based on scientific evidence, is available in Annex 3.

4.1. PHYSICAL INTERVENTIONS

4.1.1. Air-drying heat treatment

Air-drying heat treatment consists of a dry air decontamination apparatus, which produces repeatable and known heating time–temperature cycles onto food surfaces (McCann *et al.*, 2006). The use of air-drying has been proposed for the decontamination of surfaces of (smaller) meat pieces but has only been examined on a laboratory scale. Beef sample surfaces inoculated with STEC O157:H7, heated at 60 °C, 75 °C, 90 °C and 100 °C using fast and slow heating rates and subsequently held at these temperatures for up to 600 s were found to have reductions of STEC O157:H7 by 4.18–6.06 $\log_{10}$ CFU/cm^2 at the higher temperatures (90 °C and 100 °C) (McCann *et al.*, 2006). However, the significant number of microorganisms that survived posed potential concerns. Osmotic and thermal protective traits of the resistant bacterial population must be investigated before this method can be proposed as a decontamination process. Air-drying heat treatment also changed meat appearance and colour, which limited the utility of this intervention to products used for catering and institutional preparations rather than for retail sales.

> *The degree of support for the use of air-drying heat treatment as a post-processing intervention specifically for the control of STEC in raw beef was low to medium.*

4.1.2. Condensing steam

Steam is an important intervention for pathogen reduction used during processing but has also been developed for use on finished meat and have been tested at a laboratory scale. Logue *et al.* (2005) used steam treatment temperatures of 55 °C, 65 °C and 75 °C on food surfaces for 10 min, 18 s, and 10 s, and found populations of inoculated STEC O157:H7 (~6 $\log_{10}$ CFU/cm^2) were reduced the most at higher temperatures (75 °C for 10 s at 38.6 KPa, 5.59 - 3.48 $\log_{10}$ CFU/cm^2). However, post-process storage conditions were also important to ensure that no re-growth of the pathogen occurred and this was best achieved through storage under vacuum at 0 °C. This study indicated that sub-atmospheric steam could have significant application in the decontamination of post-fabrication meat primals immediately prior to packaging (Logue *et al.*, 2005).

> *The degree of support for the use of condensed steam as a post-processing intervention specifically for the control of STEC in raw beef was low.*

4.1.3. Hot water

Hot water treatment was examined as an intervention to minimize the risk of internalizing inoculated STEC O157:H7 on the surfaces of sub-primal cuts undergoing blade tenderization or moisture enhancement (Heller *et al.*, 2007). Evaluated under laboratory conditions, round pieces cut from sub-primals were inoculated with a STEC O157:H7 cocktail at 4.2 $\log_{10}$ CFU/100 cm^2. Application of hot water (82 °C for 20 s) sprayed onto the surface, resulted in 1.0 $\log_{10}$ CFU/100 cm^2 reductions in STEC O157:H7 levels. However, another study showed that use of hot water treatments (82 °C, aerobically or anaerobically [559 mm/Hg vacuum] for 3 min in a tumbler) of beef trimmings before grinding did not reduce any microorganism populations (Stivarius *et al.*, 2002).

> *The degree of support for the use of hot water treatment as a post-processing intervention specifically for the control of STEC in raw beef (sub-primals) was low.*

4.1.4. Surface trimming

Trimming of beef carcasses by slaughterhouse operators removes visible surface contamination and is effective at reducing STEC O157:H7 on sub-primals under lab conditions (Heller *et al.*, 2007). When an STEC O157:H7 cocktail (4.2 $\log_{10}$ CFU/100 cm^2) was inoculated onto round pieces cut from sub-primal meats, by trimming away the external surface with a sterile knife resulted in 1.1 $\log_{10}$ CFU/100 cm^2 reduction in inoculated STEC O157:H7 levels (Heller *et al.*, 2007). Another trial conducted under laboratory conditions also found that full-surface trimming (removal of 5 mm of the dorsal and ventral surfaces) and partial-surface trimming (removal of 5 mm from the dorsal surface only) of sub-primals, significantly decreased STEC O157:H7 levels by more than 2 $\log_{10}$ CFU/cm^2 (Lemmons *et al.*, 2011). The need to use sterile equipment for trimming poses some limitations to implementation at many locations.

> *The degree of support for the use of surface trimming as a post-processing intervention specifically for the control of STEC in raw beef (primals) was low to medium.*

4.1.5. Dry chilled ageing

Dry ageing of carcasses at refrigeration temperatures was the most common intervention used for the reduction of STEC O157:H7 in small processing plants (Tittor *et al.*, 2011). A survey found dry chilled and aged meat samples, which were suspended in refrigerators/chillers (3 °C) with an air velocity of 0.25 m/s and a relative humidity of 80 percent, resulted in a reduction of 4 $\log_{10}$ CFU/cm^2

on day 28 of storage (Tittor *et al.*, 2011). Many practices commonly used by large facilities to control pathogens and reduce faecal contamination are difficult to implement for small and very small plants. Dry chilled ageing is a critical control point and could be used as a potential intervention for small processing plants (Tittor *et al.*, 2011).

> *The degree of support for dry ageing of carcasses at refrigeration temperatures as a post-processing intervention specifically for the control of STEC in raw beef was low to medium.*

4.1.6. High pressure processing (HPP)

High pressure processing (HPP) is a safe and effective non-thermal processing method that improves the microbial safety of fresh ground beef (Zhou, Karwe and Matthews, 2016). HPP could be applied to packaged commodities, thus eliminating the potential for microbial survival and post-packaging contamination. The effect of single- and multiple-cycle HPP treatments on the survival of STEC O157:H7 in ground beef was investigated (Morales *et al.*, 2008). When HPP was applied to ground beef at 450 MPa for 15 min at refrigeration temperature (4 °C to 7 °C), more than 5 $\log_{10}$ CFU/g reduction in populations of a cocktail of O26:H11, O45:H2, O103:H2, O111:NM, O121:H19, O145 and O157:H7 serotypes was observed (Hsu *et al.*, 2015).

Sensory qualities of HPP-treated products remained unchanged so there is greater consumer acceptance than irradiated foods (Doona and Feeherry, 2007). Some research suggested that ground beef patties subjected to HPP were drier and less flavourful compared to untreated patties (Hayes *et al.*, 2014). HPP treatment retains many of the fresh qualities of the commodity. It also denatures enzymes, extends shelf-life, and reduces the need for preservatives, without significantly altering organoleptic qualities.

Combining vacuum-packaged and HPP treatment reduced STEC O157:H7 levels in ground beef by 3 $\log_{10}$ CFU/g and produced substantial sublethal injury in the surviving STEC population that resulted in further reductions in levels during frozen storage (Black *et al.*, 2010). Vacuum-packaged and HPP treatments (four, 60 s cycles, 400 MPa, 17 °C) produced > 2.0 $\log_{10}$ CFU/g reductions of *E. coli* DH5α and the major seven (O103, O111, O26, O145, O121, O45, O157:H7) STEC serogroups (Jiang *et al.*, 2015). The colour and texture of ground beef patties exhibited significant changes when more severe HPP and vacuum treatments were applied.

Refrigerated or frozen storage of HPP-processed ground beef served as an additional intervention to limit the survival and recovery of STEC O157:H7

(Black *et al.*, 2010). Combining HPP with 24 h storage of ground beef at 4 °C or at -20 °C achieved a 5 $\log_{10}$ CFU/g reduction in STEC O157:H7 populations. HPP treatment with cold storage regardless of whether the pressure applied was cyclic or static, had no significant effect on the colour of ground beef (Zhou *et al.*, 2016).

> *The degree of support for the use of HPP treatment alone or in combination with other interventions such as vacuum packaging, refrigerated or frozen storage as a post-processing intervention specifically for the control of STEC in raw beef was medium.*

4.1.7. Irradiation

The primary purpose of food irradiation is to eliminate microbial pathogens and improve food safety. Irradiation is accomplished using carefully controlled doses of ionizing radiation for a short time. One of the most common energy sources used in food irradiation is the electron beam (eBeam), which consists of highly energetic electrons, generated from commercial electricity rather than a radioactive source and it does not penetrate deeply into the product. Another common energy source is gamma radiation produced by radioactive Cobalt60 or Cesium137 which penetrate deeply into the product. Both eBeam and gamma irradiation have been approved by many different countries for non-thermal processing of foods.

The potential of eBeam to control STEC O157:H7 populations was examined using meat samples inoculated with STEC O157:H7 at levels ranging from 3–6 $\log_{10}$ CFU/cm^2. A low 1-kGy dose of eBeam radiation reduced STEC O157:H7 by at least 4 $\log_{10}$ CFU/cm^2. The impact of eBeam on organoleptic quality was assessed using flank steak and ground beef and showed that the sensory qualities of the products were not affected by the 1 KGy dose (Arthur *et al.*, 2005).

Similarly, a 1 KGy dose effected a 4 $\log_{10}$ CFU/cm^2 reduction of STEC O157:H7 and a 3.9–4.5 $\log_{10}$ CFU/cm^2 reduction was observed using a cocktail of other STEC strains (Kundu *et al.*, 2014).

Several studies examined the effects of low dose gamma irradiation to control STEC O157:H7 and other STEC. One study found that a 2.5 KGy dose reduced STEC O157:H7 populations seeded onto meat trim by 5 $\log_{10}$ CFU/g and there were no sensory changes on the product until the dosage was increased to 5 KGy (de la Paz Xavier *et al.*, 2014). Cap *et al.* 2020, used gamma irradiation on five STEC strains (O26, O103, O111, O145 and O157) seeded at 7 $\log_{10}$ CFU/g on beef trim and found that both a low (0.5 KGy) and a high (2.0 KGy) dose reduced STEC levels by 1.5 and > 5 $\log_{10}$ CFU/g, respectively. The effect of gamma irradiation on 40 STEC strains seeded onto lean ground beef was examined and results showed that the D10 value, defined as the dose needed to reduce levels by 1 $\log_{10}$

(90 percent), ranged from 0.16–0.48 kGy, with a mean of 0.31 kGy for the 40 STEC isolates (Sommers *et al.*, 2015).

These studies illustrated that both eBeam and gamma irradiation are very effective at reducing levels of STEC and STEC O157:H7. These interventions may be applicable to meats prior to the mechanical tenderization to minimize internalization of pathogens into meat or for treatment of trim prior to grinding. However, installing an irradiation system in a plant is costly, brings up security and safety concerns, and consumer perception and reluctance to buy irradiated foods may also affect the marketability of the product.

> *The degree of support for the use of eBeam and gamma irradiation as a post-processing intervention specifically for the control of STEC in raw beef was medium.*

Irradiation and organic acids

The potential of using high (HDI-2 KGy) and low (LDI-0.5 KGy) dose gamma irradiation, along with lactic acid (LA-5 percent), caprylic acid (CA-0.04 percent) or combinations of these, were evaluated as intervention measures to control five STEC strains inoculated at 7 $\log_{10}$ CFU/g on beef trim. Low dose gamma irradiation alone or with CA caused a 1.4 $\log_{10}$ reduction. Low dose gamma irradiation with LA showed a 1.7 $\log_{10}$ reduction. But the most effective treatment was HDI, which gave a > 5 $\log_{10}$ reduction. Minimal changes in meat quality parameters and sensory factors were noted with all treatments, except for LDI + LA (Cap *et al.*, 2020). Li *et al.* (2015) examined samples that were treated with 5 percent LA at 55 °C, were aerobically or vacuum packed, and kept at 4 °C. Irradiation with 1 KGy reduced STEC by 4.5 $\log_{10}$ and the addition of LA did not further reduce STEC levels. Studies showed no additional benefits of combining LA and CA with irradiation, regardless of irradiation dose in fresh beef, yet there was some benefit in frozen product.

> *The degree of support for the use of eBeam and gamma irradiation in combinaion with organic acids as a post-processing intervention specifically for the control of STEC in raw beef was medium.*

Irradiation and packaging

The effectiveness of controlling STEC O157:H7 in ground beef by combining eBeam and vacuum or modified atmosphere packaging (MAP) was examined using an inoculated mix of 5 STEC O157:H7 strains (5 $\log_{10}$ CFU/g) in ground beef patties packaged in 99.6 percent CO_2 and 0.4 percent CO or in vacuum. Patty packages that were irradiated with eBeam at 0.5, 1.0 or 1.5 KGy, showed $\log_{10}$ CFU reductions of 0.5 to 0.7, 1.0 to 2.2 and 3.0 to 3.3, respectively. The D10-values for STEC O157:H7 was similar in vacuum (0.47 ± 0.02 kGy) or in MAP (0.50 ± 0.02 kGy) and irradiated packages stored at 4 °C for 6 weeks showed no bacterial growth. However, storage at 25 °C showed growth in the vacuum packaged samples but not in MAP, suggesting that MAP was more effective than vacuum in controlling microbial growth post irradiation (Kudra et al., 2011).

The effect of combining gamma irradiation and MAP was examined with raw meat ball samples seeded with 6 $\log_{10}$ CFU/g of STEC O157:H7 and packed in MAP (3 percent O_2 and 50 percent CO_2 and 47 percent N_2) or packaged aerobically and irradiated at 0.75, 1.5, and 3 kGy before storage at 4 °C for 21 days. The D10 value for STEC O157:H7 was 0.24 KGy and it was totally inactivated by 1.5 KGy. In the aerobic packages, irradiation caused significant loss of product colour and sensory quality, but not in the MAP packages, suggesting that MAP better inhibited irradiation-induced quality degradations during the 21-day storage (Gunes et al., 2011).

The degree of support for the use of eBeam and gamma irradiation in combinaion with MAP as a post-processing intervention specifically for the control of STEC in raw beef was medium.

4.2. CHEMICAL INTERVENTIONS

4.2.1. Organic acids

Organic acids, particularly lactic acid, have a long history of use as food preservatives as well as decontamination treatments for foods, including meat. Lactic acid is probably the most widely used organic acid for meat decontamination and is already approved or in use in a number of countries. Several studies have examined the efficacy of lactic acid, although relatively few have specifically considered its efficacy on STEC O157:H7 or non-O157 STEC. EFSA (2011) reviewed studies looking at the efficacy of lactic acid treatments at a variety of concentrations for decontamination of beef carcasses, beef cuts and trimmings. For the studies which examined STEC on beef cuts and trimmings, there were reductions of 0.1 to 1.4 $\log_{10}$ CFU/g for beef cuts (Echeverry et al., 2009) and 1.1 to 2.3 $\log_{10}$ CFU/g for trimmings (Harris et al., 2006) as compared to untreated controls. Wolf et al. (2012) compared the effectiveness of 4.4 percent lactic acid dip or spray application on beef trim and in ground beef and found that dip application was more effective than spray, and decreased the levels of STEC O157:H7 by 0.91 to 1.41 $\log_{10}$ CFU/g, and non-O157 STEC were decreased by 0.48–0.82 $\log_{10}$ CFU/g.

In challenge studies that examined the use of lactic acid to decontaminate sub-primals, trimmings and cheek meat, reductions ranged from 0.2 to 2.8 $\log_{10}$ CFU for STEC O157:H7 and 0.2-3.4 $\log_{10}$ CFU for NTS E. coli (Antic, 2018).

As previously noted in Section 3.4.3, the efficacy of lactic acid treatment varied widely and was dependent on lactic acid concentration, how it was applied, the length of application, temperature and the microbial load on the meat surfaces. Other factors such as the inoculum level, STEC strain(s) used, and the recovery methods used in the studies were also impactful. In practice, lactic acid may be used in combination with other chemical or physical treatments, such as hot water or vacuum/modified atmosphere storage, which add a further level of complexity to understanding the efficacy of each treatment applied individually versus in combination.

> *The degree of support for the use of organic acids on sub-primals, trim, and cheek meat as a post-processing intervention specifically for the control of STEC in raw beef was low to medium.*

4.2.2. Other chemical treatments

A wide range of chemical treatments have been explored for the decontamination of beef, with some chemicals used individually or in various combinations. Kalchayanand *et al.* (2015) examined the efficacy of hypobromous acid, neutral acidified sodium chlorite and two citric acid-based antimicrobial compounds against strains of seven STEC serogroups (i.e. O26, O45, O103, O111, O121, O145 and O157). The chemicals when applied as spray treatments on the surface of STEC-inoculated pre-rigor beef flank at 4 °C, resulted in reductions in STEC populations of 0.7–2.0 $\log_{10}$ CFU/cm^2 after treatment and 1.2 to 2.3 $\log_{10}$ CFU/cm^2 after 48 h at 4 °C. No differences were observed in the efficacy of the four antimicrobial compounds between the strains of STEC O157:H7 and the six non-O157 STEC serovars. However, when the seven STEC strains were inoculated at low concentrations, none of the four antimicrobial treatments resulted in the complete elimination of STEC.

Muriana *et al.* (2019) examined the effectiveness of 14 different commercially-available chemical treatments with a wide range of pH values (0.8–13.1). The chemicals were sprayed at commercially recommended concentrations onto lean beef wafers (cut from cores of beef sub-primals) inoculated with a four-strain cocktail of STEC O157:H7. Reductions achieved ranged from 0.1–1.18 $\log_{10}$ CFU/cm^2 after 1 h, 0.44–2.07 $\log_{10}$ CFU/cm^2 after 1 day, and 0.37–3.61 $\log_{10}$ CFU/cm^2 after 7 days of storage. In a separate experiment, inoculated beef cores from sub-primals were subjected to antimicrobial organic acid treatments prior to blade tenderization. None of the antimicrobial treatments eliminated STEC O157:H7 post-tenderization, but there was a significant reduction in the number of positive samples when the antimicrobial treatments were used before tenderization.

Scott-Bullard *et al.* (2017) investigated the efficacy of a sulfuric acid-sodium sulfate mix against a mixture of five strains of STEC O157:H7 and 12 strains of non-O157 STEC on pre-rigor beef tissue. Treatments lowered STEC populations by 0.6–1.5 $\log_{10}$ CFU/cm^2, depending on the inoculum type and the recovery culture medium used. Similar results were obtained with samples seeded with NTS *E. coli*, supporting its suitability as a surrogate organism for STEC in efficacy validation studies of sulfuric acid-sodium sulfate in beef plants.

> *The degree of support for the use of other chemical treatments on sub-primals, trim, and cheek meat as a post-processing intervention specifically for the control of STEC in raw beef was low to medium.*

4.2.3. Ozone

Ozone is a powerful oxidising agent that exhibits antimicrobial activity against a wide range of microorganisms by damaging their cell walls and membranes leading to lysis. Ozone has been considered for use on a wide range of foods of both animal and non-animal origin with varying degrees of success, but it is most effective as an antimicrobial treatment for low pH foods (such as fruits) due to the lower decomposition of ozone in those conditions (Kumar and Sabikhi, 2019). Ozone treatments can also impact organoleptic properties of the meat, such as colour and aroma. Coll Cárdenas *et al.* (2011) found that treatment of beef primals with 72 ppm of gaseous ozone at 0 °C and 4 °C for 3 h reduced *E. coli* counts by 0.6–1.0 $\log_{10}$ CFU/g with the reduction being slightly higher at 0 °C than at 4 °C. Compared to the 3 h exposure, a longer exposure time (24 h) to ozone resulted in a greater reduction in *E. coli* counts (0.7–2.0 $\log_{10}$ CFU/g) but had a significant effect on surface colour of the beef due to lipid oxidation.

Novak and Yuan (2003) examined the effect of 3 ppm aqueous ozone treatment on a cocktail of three STEC O157:H7 strains which were applied to irradiated, sterilised beef. Ozone treatment resulted in a reduction of 0.85 $\log_{10}$ CFU/g and produced some ultrastructural changes but did not alter the visual appearance of the beef (Novac and Yuan, 2003). McMillin and Michel (2000) examined high concentrations of ozone (500, 3500 and 5000 ppm) to treat *E. coli* inoculated into minced beef and obtained up to 2 $\log_{10}$ CFU/g reductions in a dose-dependent manner; and no organoleptic changes were reported (McMillin and Michel, 2000).

> *The degree of support for the use of ozon on primals and ground beef as a post-processing intervention specifically for the control of STEC in raw beef was low.*

4.2.4. Lactoferricin B

Lactoferricin B is an antimicrobial peptide derived from acid-pepsin digestion of bovine lactoferrin and it is bactericidal to a wide range of bacteria. The antimicrobial properties of lactoferricin B are reduced at acid pH or completely inhibited by the addition of 5 percent cow's milk (Jones *et al.*, 1994). Venkitanarayanan, Zhao and Doyle (1999) investigated the effect of Lactoferricin B on coarsely ground top round beef steak which were inoculated with a five-strain mixture of STEC O157:H7. The ground beef samples were treated with lactoferricin B (100 µg/g), mixed and stored at 4 °C and 10 °C for 3 days. Lactoferricin B significantly reduced STEC O157:H7 counts by approximately 0.8 $\log_{10}$ CFU/g at both storage temperatures. However, there was no significant impact of lactoferricin B on total aerobic plate counts (Venkitanarayanan, Zhao and Doyle, 1999). Reductions in STEC O157:H7 achieved in ground beef were much lower than the 5 $\log_{10}$-cycle reductions in 1 percent peptone water at 37 °C that was reported by Shin *et al.* (1998). The study by Venkitanarayanan, Zhao and Doyle (1999) used a 10-fold lower concentration of lactoferricin B and the storage temperatures used were also much lower to simulate conditions at retail and in the home. There was no sensory analysis of the treated ground beef, so the effects of lactoferricin B on product quality remain unknown.

The degree of support for the use of lactoferricin B in ground beef as a post-processing intervention specifically for the control of STEC in raw beef was low.

4.2.5. Essential oils

A wide range of plant essential oils have a long history of antimicrobial activity particularly in *in-vitro* studies (Quinto *et al.*, 2019; Valdivieso-Ugarte *et al.*, 2019). Antimicrobial activity of essential oils has been examined in foods including meat and meat products, but the concentrations of essential oils required to be effective can result in organoleptic changes. Solomakos *et al.* (2008) examined the antimicrobial effect of thyme essential oil, nisin (a bacteriocin) or their combination against two strains of STEC O157:H7 in minced beef during refrigerated storage. Thyme oil (0.6 percent) was added to samples of minced meat inoculated with STEC O157:H7, mixed and stored at 4 °C or 10 °C for up to 12 days. There was no significant reduction in STEC O157:H7 counts with storage at 4 °C but, approximately a 1 $\log_{10}$ CFU reduction in counts were obtained at 10 °C after 2 days. The addition of 500 or 1000 IU/g of nisin along with 0.6 percent thyme oil reduced STEC O157:H7 counts by 1.0 $\log_{10}$ CFU/g at 4 °C, compared with either thyme oil or nisin alone. Sensory evaluation revealed that the organoleptic properties of 0.6 percent thyme oil treated minced beef was acceptable, however this may not be the case with other essential oils and therefore, each will have to be examined individually.

The degree of support for the use of essential oil on muscle tissue and ground beef as a post-processing intervention specifically for the control of STEC in raw beef was low.

4.3. BIOLOGICAL INTERVENTIONS

4.3.1. Bacteriophages

The potential of using bacteriophages to control STEC at post-processing of meats has been explored. A bacteriophage specific to STEC O157:H7 and tested to lyse STEC O157:H7 strains, reduced STEC O157:H7 levels by 2.7 $\log_{10}$ CFU on meat incubated at 37 °C (Hudson *et al.*, 2013). Phage treatment at 10 $\log_{10}$ PFU on enteropathogenic *E. coli* (EPEC) and STEC, reduced STEC by ~0.77 logs in 3 h, and 1.15 $\log_{10}$ after 6 h at 24 °C (Tomat *et al.*, 2013). Bacteriophage insensitive mutants (BIM) emerged at low frequency, although the use of a phage cocktail could potentially limit this development. A cocktail of six phages specific to *E. coli*, EPEC and STEC reduced STEC populations by 3–3.8 $\log_{10}$ at 37 °C, with effectiveness being time- and temperature-dependent (Tomat *et al.*, 2018). A different phage cocktail tested reduced STEC levels by 0.48 and 1.97 $\log_{10}$ at 4 °C and 24 °C, respectively, over 24 h (Hong, Pan and Ebner, 2014).

Bacteriophage do not impart sensory changes to beef. Most of the studies have been performed on small pieces of meat or packages, so it is uncertain if and how this technology can be scaled up for use in the production plant. Other concerns include the lengthy incubation time and the temperatures of 24 °C to 37 °C needed for optimal phage lytic activity, which are not well suited for conditions in the post-processing of meats.

The degree of support for the use of Bacteriophage treatment in ground beef as a post-processing intervention specifically for the control of STEC in raw beef was low.

4.3.2. Lactic acid bacteria (LAB)

Lactic acid bacteria (LAB) are antagonistic to the growth of other bacteria, including STEC. The effectiveness of LAB treatment on STEC and STEC O157:H7 was evaluated. After refrigerated vacuum ageing of beef strips for 14–28 days, LAB treatment reduced the level of a STEC cocktail by 0.4 $\log_{10}$ CFU/cm^2 (Kirsch *et al.*, 2017). After 3 days of storage, LAB-treated ground beef showed STEC O157:H7 population had reduced by 2 $\log_{10}$ at 5 °C and LAB caused no sensory changes on the meat (Smith *et al.*, 2005). Treatment with LAB reduced STEC and STEC O157:H7 populations on beef, however the lengthy time required for LAB to be effective limits its use mostly to the ageing and storage phases of the final product. Impacts of LAB on shelf-life, stability and organoleptic properties of the product have not been clearly elucidated to support the use of LAB to reduce STEC in beef.

The degree of support for the use of LAB in ground beef as a post-processing intervention specifically for the control of STEC in raw beef was low.

4.3.3. Colicin

Colicins are antimicrobial proteins produced by *E. coli* that kill other *E. coli*, including STEC O26, O45, O103, O111, O121, O145, O157 and O104:H4. Pieces of pork steak drip-inoculated with STEC and a solution of colicin M (3 mg/kg) and colicin E7 (1 mg/kg) showed that STEC levels were reduced by 2.3 and 2.7 $\log_{10}$ CFU/cm^2 after 1 h and 24 h, respectively (Schulz *et al.*, 2015).

E. coli colicins are costly to produce and available only in small amounts. However, recombinant colicin can now be made in yeast, tobacco, spinach and bean plant tissues in quantities that are feasible for commercial application. Colicins do not appear to effect sensory changes in the product, but more extensive studies are needed to better evaluate the use of colicins to control STEC in meats.

The degree of support for the use of colicin in sub-primals as a post-processing intervention specifically for the control of STEC in raw beef was low.

References

Antic, D. 2018. A critical literature review to assess the significance of intervention methods to reduce the microbiological load on beef through primary production. University of Liverpool. Food Standards Agency Project FS301044. www.food.gov. uk/print/pdf/node/4506

Arthur, T. M., Wheeler, T. L., Shackelford, S. D., Bosilevac, J. M., Nou, X. & Koomaraie, M. 2005. Effects of low-dose, low-penetration electron beam irradiation of chilled beef carcass surface cuts on *Escherichia coli* O157:H7 and meat quality. *Journal of Food Protection*, 68: 666–672. doi: 10.4315/0362-028x-68.4.666

Black, E. P., Hirneisen, K. A., Hoover, D. G. & Kniel, K. E. 2010. Fate of *Escherichia coli* O157:H7 in ground beef following high-pressure processing and freezing. *Journal of Applied Microbiology*, 108: 1352–1360. doi: 10.1111/j.1365-2672.2009.04532.x

Cap, M., Cingolani, C., Lires, C., Mozgovoj, M., Soteras, T., Sucari, A., Gentiluomo, J., Descalzo, A., Grigioni, G., Signorini, M., Horak, C., Vaudagna, S. & Leotta, G. 2020. Combination of organic acids and low-dose gamma irradiation as antimicrobial treatment to inactivate Shiga toxin-producing *Escherichia coli* inoculated in beef trimmings: Lack of benefits in relation to single treatments. *PLOS ONE*, 15(3). doi: 0.1371/journal.pone.0230812

Coll Cárdenas, F., Andrés, S., Giannuzzi, L. & Zaritzky, N. 2011. Antimicrobial action and effects on beef quality attributes of a gaseous ozone treatment at refrigeration temperatures. *Food Control*, 22: 1442–1447. doi: 10.1016/j.foodcont.2011.03006

Currie, A., Honish, L., Cutler, J., Locas, A., Lavoie, M. C., Gaulin, C., Galanis, E., Tschetter, L., Chui, L., Taylor, M., Jamieson, F., Gilmour, M., Ng, C., Mutti, S., Mah, V., Hamel, M., Martinez, A., Buenaventura, E., Hoang, L., Pacagnella, A., Ramsay, D., Bekal, S., Coetze, K., Berry, C. & Farber, J. 2019. Outbreak of *Escherichia coli* O157:H7 infections linked to mechanically tenderized beef and the largest beef recall in Canada, 2012. *Journal of Food Protection*, 82: 1532–1538. doi: 10.4315/0362-028x.jfp-19-005

Doona, C. J. & Feeherry, F. E. 2007. *High pressure processing of foods*. IFT Press/Blackwell Publishing, Ames, IA.

Duffy G., Brien, S. O., Carney, E., Butler, F., Cummins, E., Nally, P., Mahon, D., Henchion, M. & Cowan, C. 2005. A quantitative risk assessment of *E. coli* O157:H7 in Irish minced beef. In: *Teagasc, The National Food Centre Research Report* No. 74. ISBN 1841704318.

Echeverry A., Brooks, J. C., Miller, M. F., Collins, J. A., Loneragan, G. H. & Brashears, M. M. 2009. Validation of intervention strategies to control *Escherichia coli* O157:H7 and *Salmonella* Typhimurium DT 104 in mechanically tenderized and brine-enhanced beef. *Journal of Food Protection*, 72: 1616–1623. doi: 10.4315/0362-028x-72.8.1616

EFSA (European Food Safety Authority). 2011. Scientific Opinion on the evaluation of the safety and efficacy of lactic acid for the removal of microbial surface contamination of beef carcasses, cuts and trimmings. *EFSA Journal*, 9: 2317–2343. doi: 10.2903/j.efsa.2011.2317

Gunes, G., Ozturk, A., Yilmaz, N. & Ozcelik, B. 2011. Maintenance of safety and quality of refrigerated ready-to-cook seasoned ground beef product (meatball) by combining gamma irradiation with modified atmosphere packaging. *Journal of Food Science*, 76: M413–M420. doi: 10.1111/j.1750-3841.2011.02244.x

Harris, K., Miller, M. F., Loneragan, G. H. & Brashears, M. M. 2006. Validation of the use of organic acids and acidified sodium chlorite to reduce *Escherichia coli* O157 and *Salmonella* Typhimurium in beef trim and ground beef in a simulated processing environment. *Journal of Food Protection*, 69: 1802–1807. doi: 10.4315/0362-028x-69.8.1802

Hayes, J. E., Raines, C. R., De Pasquale, D. A. & Cutter, C. N. 2014. Consumer acceptability of high hydrostatic pressure (HHP)-treated ground beef patties. *LWT – Food Science and Technology*, 56(1): 207–210. doi: 10.1016/j.lwt.2013.11.014

Heller, C. E., Scanga, J. A., Sofos, J. N., Belk, K. E., Warren-Serna, W., Bellinger, G. R., Bacon, R. T., Rossman, M. L. & Smith, G. C. 2007. Decontamination of beef subprimal cuts intended for blade tenderization or moisture enhancement. *Journal of Food Protection*, 70: 1174–1180. doi: 10.4315/0362-028x-70.5.1174

Hong, Y., Pan, Y. & Ebner, P. D. 2014. *Meat Science* and Muscle Biology Symposium: Development of bacteriophage treatments to reduce *Escherichia coli* O157:H7 contamination of beef products and produce. *Journal of Animal Science*, 92: 1366–1377. doi: 10.2527/jas.2013-7272

Hsu, H., Sheen, S., Sites, J., Cassidy, J., Scullen, B. & Sommers, C. 2015. Effect of high pressure processing on the survival of Shiga toxin-producing *Escherichia coli* (Big Six vs. O157:H7) in ground beef. *Food Microbiology*, 48: 1–7. doi: 10.1016/j. fm.2014.12.002

Hudson, J. A., Billington, C., Cornelius, A. J., Wilson, T., On, S. L. W., Premaratine, A. & King, N. J. 2013. Use of a bacteriophage to inactivate *Escherichia coli* O157:H7 on beef. *Food Microbiology*, 36: 14–21. doi: 10.1016/j.fm.2013.03.006

Jiang, Y., Scheinberg, J. A., Senevirathne, R. & Cutter, C. N. 2015. The efficacy of short and repeated high-pressure processing treatments on the reduction of non-O157:H7 Shiga-toxin producing *Escherichia coli* in ground beef patties. *Meat Science*, 102: 22–26. doi: 10.1016/j.meatsci.2014.12.001

Jones, E. M., Smart, A., Bloomberg, G., Burgess, L. & Millar, M. R. 1994. Lactoferricin, a new antimicrobial peptide. *Journal of Applied Bacteriology*, 77: 208–214. doi: 10.1111/j.1365-2672.1994.tb03065.x

Kalchayanand, N., Arthur, T. M., Bosilevac, J. M., Schmidt, J. F., Wang, R., Shackelford, S. & Wheeler, T. L. 2015. Efficacy of antimicrobial compounds on surface decontamination of seven Shiga toxin–producing *Escherichia coli* and *Salmonella* inoculated onto fresh beef. *Journal of Food Protection*, 78: 503–510. doi: 10.4315/0362-028x.jfp-14-268

Khadre, M. A., Yousef, A. E. & Kim, J. G. 2001. Microbiological aspects of ozone applications in food: a review. *Journal of Food Science*, 66, 1242–1252.

Kirsch, K.R., Tolen, T.N., Hudson, J.C., Castillo, A., Griffin, D. & Taylor, T.M. 2017. Effectiveness of a commercial lactic acid bacteria intervention applied to inhibit Shiga toxin-producing *Escherichia coli* on refrigerated vacuum-aged beef. International *Journal of Food Science*, 8070515. doi: 10.1155/2017/8070515

Kudra, L., Sebranek, J. G., Dickson, J. S., Mendonca, A. F., Larson, E. M., Jackson-Davis A. L. & Lu, Z. 2011. Effects of vacuum or modified atmosphere packaging in combination with irradiation for control of *Escherichia coli* O157:H7 in ground beef patties. *Journal of Food Protection*, 74: 2018–2023. doi: 10.4315/0362-028X.JFP-11-289

Kumar, C. T. M. & Sabikhi, L. 2019. Ozone application in food processing. In O. Chauhan, eds. Chapter 11: Non-thermal processing of foods, Boca Raton: CRC Press. doi: 10.1201/b22017

Kundu, D., Gill, A., Lui, C., Goswami, N. & Holley, R. 2014. Use of low dose e-beam irradiation to reduce *E. coli* O157:H7, non-O157 (VTEC) *E. coli* and *Salmonella* viability on meat surfaces. *Meat Science*, 96: 413–418. doi: 10.1016/j.meatsci.2013.07.034

Lemmons, J. L., Lucia, L. M., Hardin, M. D., Savell, J. W. & Harris, K. B. 2011. Evaluation of *Escherichia coli* O157:H7 translocation and decontamination for beef vacuum-packaged subprimals destined for nonintact use. *Journal of Food Protection*, 74: 1048–1053. doi: 10.4315/0362-028X.JFP-10-535

Li, S. L., Kundu, D. & Holley, R. A. 2015. Use of lactic acid with electron beam irradiation for control of *Escherichia coli* O157:H7, non-O157 VTEC *E. coli*, and *Salmonella* serovars on fresh and frozen beef. *Food Microbiology*, 46: 34–39. doi: 10.1016/j.fm.2014.06.018

Logue, C. M., Sheridan, J.J. & Harrington, D. 2005. Studies of steam decontamination of beef inoculated with *Escherichia coli* O157:H7 and its effect on subsequent storage. *Journal of Applied Microbiology*, 98: 741–751. doi: 10.1111/j.1365-2672.2004.02511.x

McCann, M. S., McGovern, A. C., McDowell, D. A., Blair, L. S. & Sheridan, J. J. 2006. Surface decontamination of beef inoculated with *Salmonella* Typhimurium DT104 or *Escherichia coli* O157:H7 using dry air in a novel heat treatment apparatus. *Journal of Applied Microbiology*, 101: 1177–1187. doi: 10.1111/j.1365-2672.2006.02988.x

McMillin, K. & Michel, M. 2000. Reduction of *E. coli* in ground beef with gaseous ozone. *Louisiana Agriculture*, 43: 35.

Morales, P., Calzada, J., Ávila, M. & Nuñez, M. 2008. Inactivation of *Escherichia coli* O157:H7 in ground beef by single-cycle and multiple-cycle high-pressure treatments. *Journal of Food Protection*, 71: 811–815. doi: 10.4315/0362-028x-71.4.811

Muriana, P., Eager, J., Wellings, B., Morgan, B., Nelson, J. & Kushwaha, K. 2019. Evaluation of antimicrobial interventions against *E. coli* O157:H7 on the surface of raw beef to reduce bacterial translocation during blade tenderization. *Foods*, 8: 80. doi: 10.3390/foods8020080

Novak, J. S. & Yuan, J. T. C. 2003. Viability of *Clostridium perfringens*, *Escherichia coli*, and *Listeria monocytogenes* surviving mild heat or aqueous ozone treatment on beef followed by heat, alkali, or salt stress. *Journal of Food Protection*, 66: 382–389. doi: 10.4315/0362-028x-66.3.382

De la Paz Xavier, M., Dauber, C., Mussio, P., Delgado, E., Maquieira, A., Soria, A., Curuchet, A., Márquez, R., Méndeza, C. & López, T. 2014. Use of mild irradiation doses to control pathogenic bacteria on meat trimmings for production of patties aiming at provoking minimal changes in quality attributes. *Meat Science*, 98: 383–391. doi: 10.1016/j.meatsci.2014.06.037

Quinto, E. J., Caro, I., Villalobos-Delgado, L. H., Mateo, J., De-Mateo-Silleras, B. & Redondo-Del-Río, M. P. 2019. Food safety through natural antimicrobials. *Antibiotics* (Basel), 8: 208. doi: 10.3390/antibiotics8040208

Schulz, S., Stephan, A., Hahn, S., Bortesi, L., Jarczowski, F., Bettmann, U., Paschke, A.K., Tuse, D., Stahl, C.H., Giritch, A. & Gleba, Y. 2015. Broad and efficient control of major foodborne pathogenic strains of *Escherichia coli* by mixtures of plant-produced colicins. *Proceedings of the National Academy of Sciences (USA)*, 112: E5454–60. doi: 10.1073/pnas.1513311112

Scott-Bullard, B. R., Geornaras, I., Delmore, R.J., Woerner, D. R., Reagan, J. O., Morgan, J. B. & Belk, K. E. 2017. Efficacy of a blend of sulfuric acid and sodium sulfate against Shiga toxin–producing *Escherichia coli*, *Salmonella*, and Non-pathogenic *Escherichia coli* Biotype I on inoculated pre-rigor beef surface tissue. *Journal of Food Protection*, 80: 1987–1992. doi: 10.4315/0362-028x.jfp-17-022

Shin, K., Yamauchi, K., Teraguchi, S., Hayasawa, H., Tomota, M., Ostuka, Y. & Yamazaki, S. 1998. Antibacterial activity of bovine lactoferrin and its peptides against enterohemorrhagic *Escherichia coli* O157:H7. *Letters in Applied Microbiology*, 26: 407–411. doi: 10.1046/j.1472–765X.1998.00358.x

Smith, L., Mann, J. E., Harris, K., Miller, M. F. & Brashears, M. M. 2005. Reduction of *Escherichia coli* O157:H7 and *Salmonella* in ground beef using lactic acid bacteria and the impact on sensory properties. *Journal of Food Protection*, 68: 1587–1592. doi: 10.4315/0362-028x-68.8.1587

Solomakos, N., Govaris, A., Koidis, P. & Botsoglou, N. 2008. The antimicrobial effect of thyme essential oil, nisin and their combination against *Escherichia coli* O157:H7 in minced beef during refrigerated storage. *Meat Science*, 80: 159–166. doi: 10.1016/j.meatsci.2007.11.014

Sommers, C., Rajkowski, K. T., Scullen, O. J., Cassidy, J., Fratamico, P. & Sheen, S. S. 2015. Inactivation of Shiga toxin-producing *Escherichia coli* in lean ground beef by gamma irradiation. *Food Microbiology*, 49: 231–234. doi: 10.1016/j.fm.2015.02.013

Stivarius, M. R., Pohlman, F. W., McElyea, K. S. & Waldroup, A. L. 2002. Effects of hot water and lactic acid treatment of beef trimmings prior to grinding on microbial, instrumental color and sensory properties of ground beef during display. *Meat Science*, 60: 327–334. doi: 10.1016/S0309-1740(01)00127-9

Tittor, A. W., Tittor, M. G., Brashears, M. M., Brooks, J. C., Garmyn, A. J. & Miller, M. F. 2011. Effects of simulated dry and wet chilling and aging of beef fat and lean tissues on the reduction of *Escherichia coli* O157:H7 and *Salmonella*. *Journal of Food Protection*, 74: 289–293. doi: 10.4315/0362-028X.JFP-10-295

Tomat, D., Migliore, L., Aquili, V., Quiberoni, A. & Balagué, C. 2013. Phage biocontrol of enteropathogenic and Shiga toxin-producing *Escherichia coli* in meat products. *Frontier in Cellular Infection Microbiology*, 3: 20. doi: 10.3389/fcimb.2013.00020

Tomat, D., Casabonne, C., Aquilia, V., Balagué, C. & Quiberoni, A. 2018. Evaluation of a novel cocktail of six lytic bacteriophages against Shiga toxin-producing *Escherichia coli* in broth, milk and meat. *Food Microbiology*, 76: 438–442. doi: 10.1016/j.fm.2018.07.006

USDA (United States of America Department of Agriculture). 2016. Ground beef and food safety. In: *USDA, Food Safety*. Cited 20 August 2022. www.fsis.usda.gov/food-safety/safe-food-handling-and-preparation/meat/ground-beef-and-food-safety

USDA. 2017. FSIS Compliance guideline for minimizing the risk of Shiga toxin-producing *Escherichia coli* (STEC) in raw beef (including veal) processing (2017 Compliance Guideline). www.fsis.usda.gov/sites/default/files/import/Compliance-Guideline-STEC-Beef-Processing.pdf

Valdivieso-Ugarte, M., Gomez-Llorente, C., Plaza-Díaz, J. & Gil, A. 2019. Antimicrobial, antioxidant, and immunomodulatory properties of essential oils: a systematic review. *Nutrients*, 11: 2786. doi: 10.3390/nu11112786.

Venkitanarayanan, K. S., Zhao, T. & Doyle, M. P. 1999. Antibacterial effect of lactoferricin B on *Escherichia coli* O157:H7 in ground beef. *Journal of Food Protection*, 62: 747–750. doi: 10.4315/0362-028x-62.7.747

Wolf, M. J., Miller, M. F., Parks, A. R., Loneragan, G. H., Garmyn, A. J., Thompson, L. D. & Brashears, M. M. 2012. Validation comparing the effectiveness of a lactic acid dip with a lactic acid spray for reducing *Escherichia coli* O157:H7, *Salmonella*, and non-O157 Shiga toxigenic *Escherichia coli* on beef trim and ground beef. *Journal of Food Protection*, 75: 1968–1973. doi: 10.4315/0362-028x.jfp-12-038

Zhou, Y., Karwe, M. V. & Matthews, K. R. 2016. Differences in inactivation of *Escherichia coli* O157:H7 strains in ground beef following repeated high-pressure processing treatments and cold storage. *Food Microbiology*, 58: 7–12. doi: 10.1016/j.fm.2016.02.010

5

Processing and post-processing control strategies for STEC in raw milk and raw milk cheeses

This section will discuss interventions to reduce STEC prevalence and concentration in raw fluid milk and raw milk cheeses from processing to post-processing, including packaging. Interventions applied to raw milk and raw milk cheeses produced from all milk-producing domestic species (e.g. dairy cattle, sheep, goat, buffalo, yak, camel, small ruminants) were briefly discussed, however, only those that applied to products of bovine and caprine origin were included in this section of the report. A summary of processing and post-processing control measures for STEC in raw milk and raw milk cheese and their degree of support rating (high, medium, low), based on scientific evidence, is available in Annex 4.

5.1. RAW MILK PROCESSING

For the pupose of this report and in accordance with the definitions included within Codex General Standards for the Use of Dairy Terms (CXS 206-1999) (FAO and WHO, 1999) and the Code of Hygienic Practices for Milk and Milk Products (CAC/RCP 57-2004) (FAO and WHO, 2009), raw milk is describe as follows:

Raw milk: Milk (defined as the normal mammary secretion of milking animals obtained from one or more milking) which has not been heated beyond 40 °C or undergone any treatment that has an equivalent effect. This definition excludes milk that has been proceeded using methods where heat treatment above 40 °C have been applied.

Some farms produce raw drinking milk for sale at the farm gate, by a vending machine, or for wider sales distribution, including via the internet. However, there is a significant body of evidence from outbreaks and from testing that depending on the animal species of origin, raw milk can be a potential source of microbiological hazards including *Campylobacter*, *Salmonella*, STEC, *Brucella melitensis*, *Mycobacterium bovis* and tick-borne encephalitis virus (EFSA, 2015). In the EFSA Scientific Opinion (2015) on the public health risks related to the consumption of raw drinking milk, STEC was recognized as a potential hazard in raw drinking milk derived from cows, sheep, goats, horses and donkeys but not from camels. While pateurization is a very effective at killing off harmful pathogens in raw milk, the processing techniques described in this section have been evaluated as alternative treatments to mitigate the presence of STEC in raw milk.

5.1.1. Bactofugation

Bactofugation separates microorganisms and spores from milk by their differences in density. It is used in the dairy industry to remove bacterial spores, but its efficiency strongly depends on viscosity, so milk must be heated to 55–60 °C to reduce viscosity prior to treatment. Bactofugation removes 90–99 percent of the bacterial spores from milk, but its efficiency in removing vegetative cells is less consistent (Euster and Jakob, 2019). Faccia *et al.* (2013) showed that bactofugation reduced *Enterobacteriaceae* counts by 72 percent and Kosikowski *et al.* (1968), used a double bactofugation procedure at 54.4 °C to remove 95 percent of the NTS *E. coli* population in raw milk. By heating the milk to the required 55–60 °C prior to bactofugation, it no longer meets the definition of raw milk. In addition, bactifugation is also limited by added equipment costs and the limited volume of milk that can be quickly and efficiently processed.

The degree of support for the use of bactofugation as a processing intervention specifically for the control of STEC in raw milk was low.

5.1.2. Microfiltration

Microfiltration typically uses ceramic membranes of 1.4 µm pore size, which provide a bacterial retention rate of 99.93–99.99 percent (Trouvé *et al.*, 1991) in raw milk. Microfiltration needs raw milk to be heated to 50–60 °C to reduce viscosity before treatment, also excluding it from the definition of raw milk. Furthermore, it only works with skimmed milk, so the cream needs to be separated, as a result, its use in treating milk intended for making cheese is uncommon. Elwell and Barbano (2006) demonstrated that microfiltration removed up to 2 $\log_{10}$ CFU/mL of aerobic bacteria from raw milk. The need to heat the milk to 50 °C prior to microfiltration is a limitation, however, studies have also shown that cold microfiltration (6°C) can remove 3–5 $\log_{10}$ CFU/mL of aerobic bacteria down to non-detectable levels (Fritsch and Moraru, 2008; Griep, Cheng and Moraru, 2018). The effectiveness of microfiltration to remove NTS *E. coli* or STEC has not been examined. The investment in equipment and operating costs and the volume of milk that can be quickly and efficiently processed are limitations to the implementation of microfiltration.

The degree of support for the use of microfiltration as a processing intervention specifically for the control of STEC in raw milk was low.

5.1.3. High pressure processing (HPP)

Experimental work using high pressure to kill bacteria and extend product shelf-life was explored at the end of 19th century by Hite (1899) who found that high pressure treatment at 600 MPa for 1 h at room temperature extended the shelf-life of raw milk by 4 days. Use of a lower pressure treatment (200 MPa) delayed spoilage by about 24 h. Researchers have explored the effects of high pressure on a wide range of microorganisms under different conditions, as well as the underlying mechanisms of inactivation of bacteria and spores (Farkas and Hoover, 2000; Balasubramaniam, Martínez-Monteagudo and Gupta, 2015; Georget *et al.*, 2015). Few studies have specifically examined the effect of HPP on the survival of STEC O157:H7 (Patterson and Kilpatrick, 1998; Kalchayanand *et al.*, 1998), and most studies have used heat processed, rather than raw milk. Patterson *et al.* (1995) found that STEC O157:H7 strains were more resistant to HPP in ultra high temperature (UHT) milk than treatment on poultry meat and there was a significant difference in pressure sensitivity of different STEC strains. Patterson and Kilpatrick (1998) found that a 15 min treatment with 400 MPa at 50 °C, above the minimum temperature for raw milk, resulted in approximately a 5.0 $\log_{10}$ CFU/mL reduction of STEC O157:H7 in UHT milk.

Studies have also investigated the use of HPP as a treatment option for human milk intended for milk banks. Viazis *et al.* (2008) observed that when human milk was treated with 400 MPa at 21–31 °C, NTS *E. coli* in peptone solution was inactivated more quickly than in human milk.

The degree of support for the use of HPP as a processing intervention specifically for the control of STEC in raw milk was low to medium.

5.1.4. Irradiation (cold pasteurization)

Food irradiation is non-thermal and has the potential to reduce pathogen load in raw milk to increase product safety. Most studies on milk have investigated the effect of either gamma or eBeam irradiation.

The use of eBeam on raw milk was evaluated using different bacteria, including STEC O157:H7. Raw milk irradiated with a 2.0 kGy dose reduced aerobic plate count from 4 $\log_{10}$ CFU/mL to below detectable limits and the D10 value for STEC O157:H7 was 0.062 kGy (Ward *et al.*, 2020). Irradiation eBeam doses of 1 to 2 kGy did not alter the organoleptic and nutritional quality of raw milk, and although the level of vitamin B2 was decreased by 31.6 percent, it was within USDA nutritional guidelines, and no lipid oxidation was detected. However, after 7 days of refrigerated storage, there was 58 percent lipid oxidation but the oxidation did not result in the development of off-odors (Ward, Kerth and Pillai, 2017).

Installing an irradiation system in a plant is costly and consumer perception and reluctance to purchase irradiated foods may also affect the marketability of the product.

> *The degree of support for the use of irradiation or cold pasteurization as a processing intervention specifically for the control of STEC in raw milk was medium.*

5.1.5. Bacteriophage

A cocktail of six phages, specific to *E. coli*, EPEC and STEC, including STEC O157:H7, reduced *E. coli* and STEC O157:H7 counts in raw milk from 2 $\log_{10}$ CFU/mL to below the detection limit after 1 day at 4 °C. However, counts of EPEC and other STEC required 7–13 days of incubation to become non-detectable, demonstrating strain to strain variation in specificity and resistance to the phage. At 24 °C, reductions by phage treatment averaged 4 $\log_{10}$ CFU/mL for STEC (Tomat *et al.*, 2018). The temperature and long incubation times required for optimal phage lytic activity are not well suited for raw milk processing conditions.

> *The degree of support for the use of bacteriophage as a processing intervention specifically for the control of STEC in raw milk was low.*

5.2. RAW MILK CHEESE PROCESSING

For the pupose of this report and in accordance with the definitions included within Codex General Standards for the Use of Dairy Terms (CXS 206-1999) (FAO and WHO, 1999) and the Code of Hygienic Practices for Milk and Milk Products (CAC/RCP 57-2004) (FAO and WHO, 2009), raw milk cheese is describe as follows:

Raw milk cheeses: Cheeses made from raw milk. For technical purposes, cheese curd might be "cooked" (i.e. processed by application of heat).

Raw milk cheeses are a potential source of STEC human infections. The ability of STEC to survive raw milk cheese production processes depends primarily on their stress-response mechanisms, which include: general, acidic, osmotic, and heat shock stress-responses in *E. coli* (Peng *et al.*, 2011). Different NTS *E. coli*, and STEC serogroups or even strains within a serogroup can exhibit high variability in response to physiological properties. For example, STEC O157:H7 clones were not significantly more acid resistant than the NTS *E. coli*, and overall, the STEC O157:H7 clonal group was not exceptionally acid resistant (Large, Walk and Whittam, 2005). However, there are strain to strain variations in acid resistance among STEC.

The cheese manufacturing process varies widely depending on the starter cultures used, the mode and extent of the acidification and salting and the conditions and duration of ripening/ageing. All of these factors, which contribute toward the ultimate composition, body, taste and texture of the cheese, profoundly impact the fate of STEC or indicator organisms within a finished cheese product. Due to the large variety of raw milk cheese types and the different technologies used in their production, lactic cheeses will be split into soft, semi-hard or hard cheese varieties where applicable; as each of these categories have different curd warming (cooking) temperatures (not to be confused with heat treatment of milk), acidification conditions and ripening periods (Annex 4).

The diverse technologies used in making raw milk cheeses will have variable effects on strains of STEC and *E. coli* in cheeses. These impacts have been studied using artificial challenge studies in laboratory scale pilot plants, which poses a clear limitation to the supporting evidence for actual production situations. Furthermore, the STEC strains were often inoculated at concentrations higher than those found in naturally contaminated milk, thereby, raising further concerns about applicability of these data.

STEC populations can grow exponentially under certain conditions during the first hours of the cheese manufacturing process. As a consequence, the quality of raw milk used is essential to ensure the safety of raw milk cheeses. Two physicochemical factors can be used to inhibit the growth of STEC during the first hours of cheese manufacture: rapid acidification and high temperature, both of which can reduce bacterial cell numbers observed during cheese ripening.

5.2.1. Milk fermentation

The microbial ecosystems of raw milk can have a protective effect or inhibit the growth of some pathogenic and spoilage microorganisms (Quigley *et al.*, 2013). Lactic acid bacteria (LAB) can be used as biocontrol agents against foodborne pathogens. LAB produce organic acids which reduce pH and also antimicrobial substances including H_2O_2, diacetyl and bacteriocins (Dal Bello *et al.*, 2010), but the evidentiary support for the use of LAB to reduce STEC in raw milk was low.

> *The degree of support for the use of LAB as starter cultures as a processing intervention specifically for the control of STEC in raw milk cheese was low.*

5.2.2. Protective cultures

Protective bacterial cultures may be used to reduce or eliminate pathogens in cheeses. However, very few protective cultures are currently marketed for cheese production, underlining the difficulty of developing effective protective cultures for the cheese industry. Additionally, the non-specific nature of their antimicrobial activity indicate that the use of protective cultures can affect the activity of desired cheese flora, starter cultures, ripening bacteria, yeasts and molds (Gensler *et al.*, 2020), which thereby, impacts the sensory quality of the product.

A cocktail of *Hafnia alvei*, *Lactobacillus plantarum* and *Lactococcus lactis* reduced STEC O26:H11 and STEC O157:H7 populations by up to 2 $\log_{10}$ CFU/g in pasteurized and raw milk cheeses when inoculated into milk at 10^2 CFU/mL (Callon, Arliguie and Montel, 2016). For cheeses made from different raw milk batches inoculated with STEC O26:H11 at very low concentrations (0.5 and 0.05 CFU/mL) that closely simulated natural contamination levels, the protective culture cocktail reduced STEC levels (average of 2.8 $\log_{10}$ CFU/g) in all cheeses. Differences in the growth and inhibition of *E. coli* in the cheeses depended on the natural microbial composition of the raw milk batches. Further research utilizing metagenomics and transcriptomic approaches should improve our understanding of the interactions between the endogenous milk microbiota and STEC contamination (Fretin *et al.*, 2020). The efficacy of protective culture appears to be strain-dependent and further experimental challenge studies using raw milk cheeses along with actual production-relevant studies are needed to evaluate the potential of using live cultures as an STEC intervention.

> *The degree of support for the use of protective cultures as a processing intervention specifically for the control of STEC in raw milk cheese was low.*

5.2.3. Bacteriophage

The effect of adding bacteriophage during fermentation in the making of cheeses has been examined. Milk samples were inoculated with *E. coli* and STEC (including STEC O157:H7) and treated with an *E. coli*-specific phage cocktail that did not inhibit starter cultures. The phage cocktail completely inactivated *E. coli* and STEC O157:H7 strains after 8 h, but only reduced other STEC strains by < 1 $\log_{10}$ CFU/mL (Tomat *et al.*, 2013). No sensory changes were associated with the use of bacteriophage; however, the time and temperature needed for optimal phage lytic activity is not well suited to the cheese making process. There is also STEC inter-strain variation in response to phage treatment, as well as concerns about the emergence of phage-resistant strains or bacteriophage insensitive mutants (BIMs).

> *The degree of support for the use of bacteriophage as a processing intervention specifically for the control of STEC in raw milk cheese was low.*

5.2.4. Acidification, salting and cooking

Rapid acidification and high temperature inhibit the growth of STEC during the early stages of cheese manufacturing. A long acidic coagulation step (pH <4.5 for 24 h) involving lactic acid production prevented growth of STEC O26, O103, O145 and O157 and reduced their concentrations to below enumeration limits. However, acidification resulting from enzymatic coagulation was not sufficient to prevent the growth of STEC when the pH did not fall below 5 (Miszczycha *et al.*, 2013). For salting, sodium chloride stress caused differences in transcriptional induction, which were associated with the survival of phenotypes of *E. coli* strains in cheese. The *E. coli* strain that lacked significant induction in the three salt-stress response genes investigated survived poorly in cheese compared to the other *E. coli* strains (Peng *et al.*, 2014).

For making hard and extra-hard cheeses, curd cooking temperatures of 53 °C or higher applied to remove water from the curd grains, also rapidly reduced STEC and *E. coli* populations (~2–4.5 $\log_{10}$ CFU/g of thermotolerant *E. coli*), even when very high bacterial levels were present before cooking (Ercolini *et al.*, 2005; Miszczycha *et al.*, 2013; Peng *et al.*, 2013b). An *E. coli* strain carrying the "locus of heat resistance" is more thermotolerant and survived curd cooking well as compared to a strain without this gene cluster. However, a thermotolerant *E. coli* strain carrying the "locus of heat resistance" was decreased by 4 $\log_{10}$ CFU/g to < LOD in 24-h-cheese, indicating that the combination of several other stresses inhibited this strain as well (Peng *et al.*, 2013b).

> *The degree of support for the use of acidification, salting and cooking as a processing intervention specifically for the control of STEC in raw milk cheese was low to medium.*

5.2.5. Ripening and ageing

The ripening step of soft cheeses can reduce STEC populations, and reductions were higher for STEC O157:H7 than for strains of STEC O26, O103 and O145 (Perrin *et al.*, 2015; Miszczycha *et al.*, 2016). However, STEC was not eliminated completely by ripening, and different STEC serogroups displayed variations in acid-resistance and survived the ripening process of soft cheeses (Montet *et al.*, 2009).

For uncooked, pressed cheeses, the duration of ripening impacted STEC survival. A short ripening period (40 days) did not promote a significant reduction in water activity (a_w), or a reduction of STEC populations. In contrast, when aged for 240 days, the a_w decreased under the minimum value of 0.95 resulting in reduction, but not total elimination of several STEC serotypes (Miszczycha *et al.*, 2013). Similarly, STEC reduction and survival was proportional to the duration of ripening in "bleu" type cheeses, where after 240 days, STEC O26, O103 and O157 were either not detected or below the enumeration limit (Miszczycha *et al.*, 2013). A combination of factors during the ripening step (e.g. temperature, low a_w of 0.898, acidic pH and the presence of antagonistic microorganisms) likely helped to reduce STEC populations in the end products.

Semi-hard cheeses are comprised of a large variety of types, so risk assessment-based challenge studies are required to examine the effects of various specific manufacturing processes on STEC. Approximate reductions of ≥ 1 $\log_{10}$ CFU/g of STEC or *E. coli* per month was observed during ripening of semi-hard cheeses (Miszczycha *et al.*, 2013; Peng *et al.*, 2013a, 2013b). Bacterial populations decreased more rapidly in the cheese core which had higher maturation temperatures as compared with the rind. It has been speculated that differences in CO_2 partial pressure in the cheese core also contributed to faster decrease of STEC or *E. coli* populations (Peng *et al.*, 2013b).

> *The degree of support for the role of ripening and ageing during cheese production as a processing intervention specifically for the control of STEC in raw milk cheese was low to medium.*

5.2.6. Cheese size

STEC or *E. coli* are rarely detected in hard cheeses after cooking at $\geq$ 53 °C for 30 min. But, STEC could still be detected in the rind zone (although only after culture enrichment), as this area cools down most rapidly compared to the core zone. The size of a cheese block, therefore, contributes to differences in temperature profile and to the survival of pathogens. During ripening, a thermotolerant *E. coli* strain was detected in a number of samples (Peng *et al.*, 2013b), including at the end of the ripening period. However, the inoculation level, size (8 kg instead of 35 kg for commercial cheese) and the conditions used (53 °C instead of up to 57 °C cooking temperature) in this model cheese production study were more favorable for the survival of *E. coli*. In commercial production, the potential of *E. coli* surviving until the end of ripening in cooked hard raw milk cheese is expected to be low (Peng *et al.*, 2013b).

Ercolini *et al.* (2005) developed a curd-cooling model where 55 °C cooking temperature was used for cooked hard cheese made from raw milk. Variation in temperature across the young cheese which was combined with a challenge test using different pathogens including STEC O157:H7, led to conclusions that pathogens in the cheese core are eliminated by cooking at 55 °C, but may survive in the more favorable thermal conditions on the crust (absent other pathogen reduction factors), as the rind cools more rapidly than does the core. Therefore, the size of a cheese block impacts the temperature profile and the survival of pathogens (e.g. the larger the cheese, the lower the ability of survival in the core of cooked cheeses). However, other factors such as acidification, long ripening period and high NaCl concentration of the crust contributed to the reduction of pathogens in this part of the cheese.

Semi-hard cheeses with cooking temperatures of 40 °C and 46 °C (which allows growth of *E. coli*) had a significantly higher decrease in *E. coli* populations in the core than in the rind (Peng *et al.*, 2013a). This has clear implications on the outcome of STEC challenge studies, hence studies should simulate as closely as possible the size of the actual product made by the industry. If the cheese size used in the challenge test is too small, the results would reflect more the conditions that STEC would experience in the cheese rind, but not those in the core of an actual cheese product.

> *The degree of support for the size fo the cheese block as a processing intervention specifically for the control of STEC in raw milk cheese was low.*

5.3. RAW MILK CHEESE POST-PROCESSING

5.3.1. Packaging

Active packaging is used in the food industry primarily to extend shelf-life, but it can also prevent pathogen growth (Yildirim *et al.*, 2018). Although not as effective against STEC O157:H7, the use of technology such as modified atmosphere packaging can retard the growth of other pathogenic bacteria (e.g. *L. monocytogenes* and *S. aureus*) in hard cheeses made with raw sheep's milk (Solomakos *et al.*, 2019).

However, the application of active packaging in cheese is still limited and further studies are needed to explore the potential and efficacy of these technologies (Speranza *et al.*, 2020; Al-Moghazy, Mahmoud and Nada, 2020).

> *The degree of support for the use fo active packaging as a post-processing intervention specifically for the control of STEC in raw milk cheese was low.*

5.3.2. Irradiation

The efficacy of eBeam to control pathogens on the surface of raw milk cheeses was evaluated using *L. monocytogenes* inoculated onto the surface of Camembert and Brie. Samples packaged under vacuum and irradiated with eBeam doses of 1.27 and 2.59 kGy controlled the growth of *L. monocytogenes*. No significant differences were noted in the sensory attributes of the Camembert samples treated with doses up to 2.59 kGy (Velasco *et al.*, 2015). For cheese slices, other technologies such as X-ray, UV-C irradiation, and pulsed-light can be applied to reduce pathogens, including STEC O157:H7 (Park and Ha, 2019; Proulx *et al.*, 2015; Ha *et al.*, 2016).

> *The degree of support for the use of irradiation as a post-processing intervention specifically for the control of STEC in raw milk cheese was medium.*

5.3.3. Bacteriophage

In a limited study, bacteriophages were found to be not very effective in reducing STEC levels in finished cheese products. The study used 16 cm^2 pieces of cheese seeded with STEC O157:H7. Treatment with phages only showed $< 0.15 \log_{10}$ CFU/g reduction, suggesting that phages were not very effective in controlling STEC in the finished cheeses (Hong, Pan and Ebner, 2014). It was suspected that perhaps the low pH of cheese (~5.5 to 6.8) inhibited phage function, but ineffectiveness of the phage treatment is most likely due to other undetermined factors.

> *The degree of support for the use of bacteriophage as a post-processing intervention specifically for the control of STEC in raw milk cheese was low.*

References

Al-Moghazy, M., Mahmoud, M. & Nada, A. A. 2020. Fabrication of cellulose-based adhesive composite as an active packaging material to extend the shelf life of cheese. *International Journal of Biologic Macromolecules*, 160: 264–275. doi: 10.1016/j. ijbiomac.2020.05.217

Balasubramaniam, V. M., Martínez-Monteagudo, S. I. & Gupta, R. 2015. Principles and application of high pressure–based technologies in the food industry. *Annual Review of Food Science and Technology*, 6: 435–462. doi: 10.1146/annurev-food-022814-015539

Callon, C., Arliguie C. & Montel, M. C. 2016. Control of Shiga toxin-producing *Escherichia coli* in cheese by dairy bacterial strains. *Food Microbiology*, 53: 63–70. doi: 10.1016/j.fm.2015.08.009

Dal Bello, B., Rantsiou, K., Bellio, A., Zeppa, G., Ambrosoli, R., Civera, T. & Cocolin, L. 2010. Microbial ecology of artisanal products from North West of Italy and antimicrobial activity of the autochthonous populations. *LWT - Food Science and Technology*, 43: 1151–1159. doi: 10.1016/j.lwt.2010.03.008

Elwell, M. W. & Barbano, D. M. 2006. Use of microfiltration to improve fluid milk quality. *Journal of Dairy Science*, 89: E10–E30. doi: 10.3168/jds.S0022-0302(06)72361-X

Ercolini, V., Fusco, G., Blaiotta, G., Sarghini, F. & Coppola, S. 2005. Response of *Escherichia coli* O157:H7, *Listeria monocytogenes, Salmonella* Typhimurium, and *Staphylococcus aureus* to the thermal stress occurring in model manufactures of grana padano cheese. *Journal of Dairy Science*, 88: 3818–3825. doi: 10.3168/jds. S0022-0302(05)73067-8

EFSA (European Food Safety Authority). 2015. Scientific opinion of the panel on Biological Hazards on the public health risks related to the consumption of raw drinking milk. *EFSA Journal*, 13(1): 3940. doi : 10.2903/j.efsa.2015.3940

Eugster, E. & Jakob, E. 2019. Pre-treatments of milk and their effect on the food safety of cheese. *Milk Science International*, 72: 45–52. ISSN 2567-9538.

Faccia, M., Mastromatteo, M., Conte, A. & Del Nobile, M. A. 2013. Influence of the milk bactofugation and natural whey culture on the microbiological and physico-chemical characteristics of mozzarella cheese. *Journal of Food Processing and Technology*, 4: 1–7. doi: 10.4172/2157-7110.1000218

FAO & WHO. 1999. Codex Alimentarius. General Standard for the Use of Dairy Terms (CXS 206-1999). Cited 20 August 2022. www.fao.org/fao-who-codexalimentarius/ sh-proxy/tr/?lnk=1&url=https%253A%252F%252Fworkspace.fao.org%252Fsites %252Fcodex%252FStandards%252FCXS%2B206-1999%252FCXS_206e.pdf

FAO & WHO. 2009. Codex Alimentarius. Code of Hygienic Practice for Milk and Milk Products (CXC 57-2004). Cited 20 August 2022. www.fao.org/fao-who-codexalimentarius/sh-proxy/en/?lnk=1&url=https%253A%252F%252Fworkspace.fao.org%252Fsites%252Fcodex%252FStandards%252FCXC%2B57-2004%252FCXC_057e.pdf

Farkas, D. F. & Hoover, D. G. 2000. High pressure processing. *Journal of Food Science*, 65: 47–64. doi: 10.1111/j.1750-3841.2000.tb00618.x

Frétin, M., Chassard, C., Delbès, C., Lavigne, R., Rifa, E., Theil, S., Fernandez, B., Laforce, P. & Callon, C. 2020. Robustness and efficacy of an inhibitory consortium against *E. coli* O26:H11 in raw milk cheeses. *Food Control*, 115: 107282. doi: 10.1016/j.foodcont.2020.107282

Fritsch, J. & Moraru, C.I. 2008. Development and optimization of a carbon dioxide-aided cold microfiltration process for the physical removal of microorganisms and somatic cells from skim milk. *Journal of Dairy Science*, 91: 3744–3760. doi: 10.3168/jds.2007-0899

Gensler, C. A., Brown, S. R. B., Aljasir, S. F. & D'Amico, D. J. 2020. Compatibility of commercially produced protective cultures with common cheesemaking cultures and their antagonistic effect on foodborne pathogens. *Journal of Food Protection*, 83: 1010–1019. doi: 10.4315/JFP-19-614

Georget, E., Sevenich, R., Reineke, K., Mathys, A., Heinz, V., Callanan, M., Rauh, C. & Knorr, D. 2015. Inactivation of microorganisms by high isostatic pressure processing in complex matrices: A review. *Innovative Food Science and Emerging Technologies*, 27: 1–14. doi: 10.1016/j.ifset.2014.10.015

Griep, E. R., Cheng, Y. & Moraru, C. I. 2018. Efficient removal of spores from skim milk using cold microfiltration: Spore size and surface property considerations. *Journal of Dairy Science*, 101: 9703–9713. doi: 10.3168/jds.2018-14888

Ha, J. W., Back, K. H., Kim, Y. H. & Kang, D. H. 2016. Efficacy of UV-C irradiation for inactivation of food-borne pathogens on sliced cheese packaged with different types and thicknesses of plastic films. *Food Microbiology*, 57: 172–177. doi: 10.1016/j.fm.2016.02.007

Health Canada. 2014. Guidance on Mandatory Labelling for Mechanically Tenderized Beef. In: *Canada.ca, Food and nutrition legislation and guidelines*. Cited 22 August 2022. www.canada.ca/content/dam/hc-sc/migration/hc-sc/fn-an/alt_formats/pdf/legislation/guide-ld/mech-tenderized-beef-boeuf-attendris-meca-eng.pdf

Hong, Y., Pan, Y. & Ebner, P. D. 2014. Development of bacteriophage treatments to reduce *Escherichia coli* O157:H7 contamination of beef products and produce. *Journal of Animal Science*, 92: 1366–1377. doi: 10.2527/jas.2013-7272

Hite, B. H. 1899. The effects of pressure in the preservation of milk. Morgantown. *Bulletin of the University of West Virginia University Agricultural Experiment Station*, 58: 15–35.

Kalchayanand, N., Sikes, A., Dunne, C. P. & Ray, B. 1998. Interaction of hydrostatic pressure, time and temperature of pressurization and pediocin AcH on inactivation of foodborne bacteria. *Journal of Food Protection*, 61: 425–431. doi: 10.4315/0362-028X-61.4.425

Kosikowski, F. V. & Fox, P. F. 1968. Low heat, hydrogen peroxide, and bactofugation treatments of milk to control coliforms in cheddar cheese. *Journal of Dairy Science*, 51: 1018–1022. doi: 10.3168/jds.S0022-0302(68)87116-4

Large, T. M., Walk, S. T. & Whittam, T. S. 2005. Variation in acid resistance among Shiga toxin-producing clones of pathogenic *Escherichia coli*. *Applied and Environmental Microbiology*, 71: 2493–2500. doi: 10.1128/AEM.71.5.2493–2500.2005

Miszczycha, S. D., Bel, N., Gay-Perret, P., Michel, V., Montel, M. C. & Sergentet-Thevenot, D. 2016. Short communication: Behavior of different Shiga toxin-producing *Escherichia coli* serotypes (O26:H11, O103:H2, O145:H28, O157:H7) during the manufacture, ripening, and storage of a white mold cheese. *Journal of Dairy Science*, 99(7): 5224–5229. doi: 10.3168/jds.2015-10803

Miszczycha, S. D., Perrin, F., Ganet, S., Jamet, E., Tenehaus-Aziza, F., Montel, M. & Theenot-Sergentet, D. 2013. Behavior of different Shiga toxin-producing *Escherichia coli* serotypes in various experimentally contaminated raw-milk cheeses. *Applied Environmental Microbiology*, 79: 150–158. doi: 10.1128/AEM.02192-12

Montet, M. P., Jamet, E., Ganet, S., Dizin, M., Miszczycha, S., Dunière, L., Thevenot, D. & Vernozy-Rozand, C. 2009. Growth and survival of acid-resistant and non-acid-resistant Shiga-toxin-producing *Escherichia coli* strains during the manufacture and ripening of camembert cheese. *International Journal of Microbiology*, 2009: 653481. doi: 10.1155/2009/653481

Park, J. S & Ha, J. W. 2019. X-ray irradiation inactivation of *Escherichia coli* O157:H7, *Salmonella enterica* Serovar Typhimurium, and *Listeria monocytogenes* on sliced cheese and its bactericidal mechanisms. *International Journal of Food Microbiology*, 289: 127–133. doi: 10.1016/j.ijfoodmicro.2018.09.011

Patterson, M. F. & Kilpatrick, D. J. 1998. The combined effect of high hydrostatic pressure and mild heat on inactivation of pathogens in milk and poultry. *Journal of Food Protection*, 61: 432–436. doi: 10.4315/0362-028X-61.4.432

Patterson, M. F., Quinn, M., Simpson, R. & Gilmour, A. 1995. Sensitivity of vegetative pathogens to high hydrostatic pressure treatment in phosphate buffered saline and foods. *Journal of Food Protection*, 58: 524–529. doi: 10.4315/0362-028X-58.5.524

Peng, S., Tasara, T., Hummerjohann, J. & Stephan, R. 2011. Overview of molecular stress response mechanisms in *Escherichia coli* contributing to survival of Shiga toxin-producing *Escherichia coli* during raw milk cheese production. *Journal of Food Protection*, 74: 849–864. doi: 10.4315/0362-028X.JFP-10-469

Peng, S., Hoffmann, W., Bockelmann, W., Hummerjohann, J., Stephan, R. & Hammer, P. 2013a. Fate of Shiga toxin-producing and generic *Escherichia coli* during production and ripening of semihard raw milk cheese. *Journal of Dairy Science*, 96: 815–23. doi: 10.3168/jds.2012-5865

Peng, S., Schafroth, K., Jakob, E., Stephan, R. & Hummerjohann, J. 2013b. Behaviour of *Escherichia coli* strains during semi-hard and hard raw milk cheese production. *International Dairy Journal*, 31: 117–120. doi: 10.1016/j.idairyj.2013.02.012

Peng, S., Stephan, R., Hummerjohann, J. & Tasara, T. 2014. Transcriptional analysis of different stress response genes in *Escherichia coli* strains subjected to sodium chloride and lactic acid stress. *FEMS Microbiology Letters*, 361: 131–137. doi: 10.1111/1574-6968.12622

Perrin, F., Tenenhaus-Aziza, F., Michel, V., Miszczycha, S., Bel, N. & Sanaa, M. 2015. Quantitative risk assessment of haemolytic and uremic syndrome linked to O157:H7 and non-O157:H7 Shiga-toxin-producing *Escherichia coli* strains in raw milk soft cheeses. *Risk Analysis*, 35: 109–128. doi: 10.1111/risa.12267

Proulx, J., Hsu, L. C., Miller, B. M., Sullivan, G., Paradis, K. & Moraru, C. I. 2015. Pulsed-light inactivation of pathogenic and spoilage bacteria on cheese surface. *Journal of Dairy Science*, 98: 5890–5898. doi: 10.3168/jds.2015-9410

Quigley, L., O'Sullivan, O., Stanton, C., Beresford, T. P., Ross, R. P., Fitzgerald, G. F., Cotter, P. D. 2013. The complex microbiota of raw milk. *FEMS Microbiol Reviews*, 37(5):664-98. doi: 10.1111/1574-6976.12030

Solomakos, N., Govari, M., Botsoglou, E. & Pexara, A. 2019. Effect of modified atmosphere packaging on physicochemical and microbiological characteristics of *Graviera Agraphon* cheese during refrigerated storage. *Journal of Dairy Research*, 86: 483–489. doi: 10.1017/S0022029919000724

Speranza, B., Liso, A., Russo, V. & Corbo, M. R. 2020. Evaluation of the potential of biofilm formation of *Bifidobacterium longum* subsp. *infantis* and *Lactobacillus reuteri* as competitive biocontrol agents against pathogenic and food spoilage bacteria. *Microorganisms*, 8: 177–191. doi: 10.3390/microorganisms8020177

Tomat, D., Casabonne, C., Aquilia, V., Balagué, C. & Quiberoni, A. 2018. Evaluation of a novel cocktail of six lytic bacteriophages against Shiga toxin-producing *Escherichia coli* in broth, milk and meat. *Food Microbiology*, 76: 43–442. doi: 10.1016/j.fm.2018.07.006

Tomat, D., Mercanti, D., Balagué, C. & Quiberoni, A. 2013. Phage biocontrol of enteropathogenic and Shiga toxin-producing *Escherichia coli* during milk fermentation. *Letters in Applied Microbiology*, 57: 3–10. doi: 10.1111/lam.12074

Trouvé, E., Maubois, J. L., Piot, M., Madec, M. N., Fauquant, J., Rouault, A., Tabard, J. & Brinkman, G. 1991. Rétention de différentes espèces microbiennes lors de l'épuration du lait par microfiltration en flux tangentiel. *Lait*, 71: 1–13. doi: 10.1051/lait:199111

Velasco, R., Ordóñez, J. A., Cambero, M. I. & Cabeza, M. C. 2015. Use of E-beam radiation to eliminate *Listeria monocytogenes* from surface mould cheese. *International Microbiology*, 18: 33–40. doi: 10.2436/20.1501.01.232

Viazis, S., Farkas, B. E. & Jaykus, L. A. 2008. Inactivation of bacterial pathogens in human milk by high-pressure processing. *Journal of Food Protection*, 71: 109–118. doi: 10.4315/0362-028X-71.1.109

Ward, L. R., Van Schaik, D., Samuel, J. & Pillai, S. D. 2020. Reduction in microbial infection risks from raw milk by electron beam technology. *Radiation Physics and Chemistry*, 168: 108567. doi: 10.1016/j.radphyschem.2019.108567

Ward, L. R., Kerth, C. R. & Pillai, D. 2017. Nutrient profiles and volatile odorous compounds of raw milk after exposure to electron beam pasteurizing doses. *Journal of Food Science*, 82: 1614–1621. doi: 10.1111/1750-3841.13763

Yildirim, S., Röcker, B., Pettersen, M., Nilsen-Nygaard, J., Ayhan, Z., Rutkaite, R., Radusin, T., Suminska, P., Marcos, B. & Coma, V. 2018. Active packaging applications for food. *Comprehensive Reviews in Food Science and Food Safety*, 17: 165–199. doi: 10.1111/1541-4337.12322

6

Primary production and processing control strategies for STEC in other animal species

6.1. SMALL RUMINANTS: PRIMARY PRODUCTION

Similar to cattle, other ruminant animals such as sheep and goats can harbour STEC, which can be transmitted through faecal contamination to food products derived from these animals (e.g. lamb, mutton, milk, and cheese). Although some STEC serotypes associated with human disease have been detected in sheep, goats and wild game they often carry a wide variety of non-O157 STEC serotypes that are well adapted for colonizing these specific animal species (Furlan *et al.*, 2019; Jacob *et al.*, 2013). While the carriage of STEC O157:H7 is less frequent, small ruminants, sheep and goats are considered to be the second most important source of human STEC infection after cattle (FAO and WHO, 2019). Because of their adaptability and low maintenance costs, sheep and goats are raised in diverse environments and conditions around the world, which creates challenges when applying STEC control interventions globally or systematically.

Relative to the amount of literature available for cattle, there is much less information on the application of interventions during primary production, processing and post-processing of products from sheep and goats. It is predicted that most GAP and GHP that are used in cattle milk or meat production will be similarly impactful in goats and sheep. For example, mixed species (sheep, deer and cattle) in the same pasture resulted in STEC transfer and increased carriage of STEC O157:H7 in sheep as well as in cattle (Section 2.2.1), and ionophore feeding did not impact on STEC O157:H7 populations in sheep – a similar result to that observed in in cattle (Section 2.4.4) (Edrington *et al.*, 2003).

While there are expectations that different farm level interventions used to reduce STEC during cattle production may also be effective in sheep and goat at the primary production stages, there are few studies that have validated their effectiveness on a commercial scale for reducing STEC in small ruminants, hence the low degree of support for most interventions in small ruminants. Interestingly, many of the studies examining interventions meant for cattle utilized (at least initially) sheep and goats in experimental infection studies with promising results.

6.1.1. Diet composition and feeding strategies

As observed in cattle, dietary components and changes in ration can influence STEC shedding in sheep. For example, sheep fed a grass hay diet shed STEC O157:H7 for twice as long a period as sheep fed a ration low in protein and digestible energy (Kudva *et al.*, 1997). Largely because of the limited amount of available data, the degree of support for the use specific diets as an intervention specifically for the control of STEC in small ruminants was low.

6.1.2. Feed additives

Probiotics

A variety of probiotics, composed of lactic acid bacteria previously described for cattle (Section 2.3.4.1), have also been used in research trials in lambs and sheep to reduce STEC O157:H7 and non-O157 STEC excretion and prevalence (Rigobelo *et al.*, 2015).

> *The degree of support for the addition of probiotics to feed as an intervention specifically for the control of STEC in small ruminants was low.*

Bacteriophage

A number of studies have investigated the potential of bacteriophage therapy to reduce STEC O157:H7 in sheep. Most of the reports have shown that bacteriophage can reduce intestinal carriage of STEC O157:H7 (Wang *et al.*, 2017; Raya *et al.*, 2011). In the rumen, reductions in STEC O157:H7 counts was not statistically different between phage treated and untreated sheep. However, bacteriophages applications significantly reduced STEC O157:H7 counts in the lower intestinal tract (up to 4 $\log_{10}$ CFU) with no adverse effects (Callaway *et al.*, 2008).

> *The degree of support for the addition of bacteriophage to feed as an intervention specifically for the control of STEC in small ruminants was low.*

Lactoferrin

The use of lactoferrin, an immunomodulatory protein found in milk, in the prevention of STEC O157:H7 colonization and excretion in sheep has also been studied. In one study, lactoferrin, administered orally to sheep, reduced STEC O157:H7 counts (up to 5 log$_{10}$ CFU) and the duration of excretion as compared to untreated controls (Yekta *et al.*, 2011).

> *The degree of support for the administration of lactoferrin as an intervention specifically for the control of STEC in small ruminants was low.*

Sodium chlorate

Feeding sodium chlorate to sheep before transport to the abattoir (a period of 24 h prior to slaughter) reduced inoculated STEC O157:H7 levels in the rumen, caecum, and colon by 1-4 log$_{10}$/g digesta (Callaway *et al.*, 2003). Results showed that treatment reduced STEC populations throughout the gut; yielding results that were very similar in scope and scale to those found from cattle (Section 2.3.4.4).

> *The degree of support for the addition of sodium chlorate to feed as an intervention specifically for the control of STEC in small ruminants was low.*

6.1.3. Vaccination

Previous experimental challenge studies of vaccinated goats with STEC O157:H7 showed promising results, but only small numbers of animals were used, so it is still experimental. In another study, goats inoculated with a STEC vaccine based on Stx2B-Tir-Stx1B-Zot protein and recombinant H7-HCP-Tir-Intimin proteins, significantly elicited Stx2b-Tir-Stx1b-Zot-specific serum IgG antibodies. When these vaccinated goats were challenged with STEC O157:H7, they showed reduced STEC O157:H7 excretion (Zhang *et al.*, 2014).

> *The degree of support for the use of vaccines as an intervention specifically for the control of STEC in small ruminants was low.*

6.1.4. Feed withdrawl prior to slaughter

In the United States of America, sheep and goats can be held off feed for up to 24 h to reduce gut fill and hide/pelt contamination; but fasting is thought to cause an increase in STEC population in sheep and goats similar to what occurs in cattle (Section 2.6.1). Withdrawing feed from sheep and goats for 12 h prior slaughter reduced NTS *E. coli*, *Enterobacteriaceae* and total coliform levels in the rumen as compared to animals that were fasted for 24 h (Gutta *et al.*, 2009; Pointon, Kiermeier and Fegan, 2012).

> *The degree of support for the use of feed withdrawal prior to slaughter as an intervention specifically for the control of STEC in small ruminants was low.*

6.2. SMALL RUMINANTS: PROCESSING (MEAT AND DAIRY)

Similar to cattle, the main sources of STEC in sheep and goat carcasses are hides/fleeces contaminated with intestinal faecal material. Controlling faecal contamination by ensuring that only clean animals are slaughtered while at the same time preventing or reducing faecal content transfer to sheep and goat carcasses during slaughter, are acceptable GHP that can also reduce STEC and other bacteria (EFSA, 2013). Wool/pelt treatments have been applied to reduce carriage of STEC into the processing facilities, but the mechanism of spraying is different than that used on cattle due to the differences between cattle hides and sheep pelts for carriage of faecal material.

When sheep carcasses are processed, the pelt is removed by a slightly different process than that of cattle. Through a process known as "fisting", which may be performed by a machine. The pelt is separated from the felt membrane which is left on the carcass to prevent shrinkage due to dehydration. In commercial sheep and goat processing, typically the carcasses are not split or ribbed as done for cattle. There were no other significant differences identified in the procedures utilized in sheep and goat meat processing, but some studies on carcasses treatments during pre-chill were evaluated.

6.2.1. Pre-chill carcass treatments

Hot water, lactic acid, and other organic acids have been used as sprays on sheep and goat carcasses to effect STEC reductions broadly similar to cattle (Section 3.4) and it is possible to achieve acceptable *E. coli* $\log_{10}$ CFU reductions. Hassan *et al.* (2015) reported a mean reduction in NTS *E. coli* of 1.1 $\log_{10}$ CFU/cm^2 on sheep and lamb carcasses treated with steam vacuum pasteurization (>82 °C for 10 s) after trimming. Another study achieved greater than 3 $\log_{10}$ CFU/cm^2 reductions of NTS *E. coli* on uninoculated and inoculated sheep carcasses by submersion or spraying hot water and steam (80 °C) or a combination of hot water and antimicrobials (Dorsa, Cutter and Siragusa, 1996). Furthermore, using antimicrobials such as lactic acid, peroxyacetic acid, a hydrochloric and citric acid blend, and levulinic acid plus 0.5 percent sodium dodecyl sulfate, on goat carcasses during slaughter and subsequent chilling was effective in reducing STEC with mean reductions between 0.47–2.26 $\log_{10}$ CFU/cm^2 (Thomas *et al.*, 2019).

The degree of support for the use of pre-chill carcasee treatments as interventions specifically for the control of STEC in small ruminants was low.

6.2.2. Dairy processing

Harvesting and processing milk from small ruminants is not markedly different from cattle, other than the physical differences in the processes depending upon the number of teats and the usual reliance on hand milking in small ruminants rather than automated systems. The use of raw milk for the production of cheese and other dairy products is similar to that for raw cows' milk.

6.3. OTHER SPECIES: PRIMARY PRODUCTION AND PROCESSING

6.3.1. Reindeer

Reindeer (*Rangifer tarandus*) are primarily used for meat production, as the use of milk from reindeer is uncommon. STEC and the presence of *stx* genes have been identified in reindeer (Zweifel *et al.*, 2017; Magwedere *et al.*, 2013; Miko *et al.*, 2020). Processing of reindeer occurs at specific reindeer slaughter facilities or by traditional field methods. Most reindeer are slaughtered at 6–7 months of age and are transported long distance via specialized trucks. The process and hygiene practices of reindeer slaughter are similar to that of cattle or sheep. In Europe, the slaughter and processing of reindeer and other farmed wild game animals is legislated (Reg. [EC] No. 852/2004 and Reg. [EC] No. 853/2004) (Laaksonen *et al.*, 2017).

GAP and GHP are suggested during reindeer primary production or meat processing; however, there was limited available evidence to support interventions specific for the control of STEC in reindeer or reindeer meat processing.

6.3.2. Yaks

Yaks (*Poephagus grunniens* or *Bos grunniens*) live at high altitude (above 3 000 m) in China, India, Nepal and other countries and are used for both meat and milk production (Rehman *et al.*, 2017; Bandyopadhyay *et al.*, 2009; Bai *et al.*, 2013). Yaks are a natural source of STEC when raised under migratory or free ranging systems. The transmission of STEC O157:H7 and non-O157 STEC from yak happens through meat, unpasteurised milk, or direct and indirect contact with humans. This occurs primarily in the nomadic herdsmen who often consume raw or undercooked yak meat, milk and milk products (e.g. "churpi", a dried and smoked hard cheese made from yak milk) (Bandyopadhyay *et al.*, 2012).

Primary production control strategies are focused on the production of meat and dairy products from the farm through transport to processing facilities, or until milk reaches the bulk tank for pasteurization or for inclusion into making raw milk cheeses.

> *GAP and GHP are suggested during yak primary production or meat and milk processing; however, there was limited available evidence to support interventions specific for the control of STEC in yaks and yak meat and dairy processing.*

6.3.3. Camel

Camel production is focused mostly in arid regions and knowledge of their production parameters is quite limited compared with other domestic animals. Camels can carry STEC, and the animals are used for both meat and raw milk production (Baschera, Cernela and Stevens, 2019; Salehi *et al.*, 2012). The life pattern of camels in the desert minimizes the contact of camels with other animal species that can be STEC reservoirs. However, during the dry season, camels can be reared along with cattle, sheep, and goats (along with exposure to birds and other wildlife) which poses increased STEC transmission risk to the camels. Camels shed STEC in their faeces (Tabatabaei *et al.*, 2013), and shedding is the highest during the wet season when feed intake (and faecal output) is increased (Adamu *et al.*, 2018).

> *GAP and GHP are suggested during camel primary production and meat and milk processing; however, there was limited available evidence to support interventions specific for the control of STEC in camels or camel meat and dairy processing.*

6.3.4. Water buffalo

Water buffalo (*Bubalus bubalis*) are typically used as draft animals and also used for milk production - primarily for making cheeses (including raw milk cheeses) and yogurt. Water buffalo are also used for meat production in Europe and Asia. Several studies have demonstrated that domestic water buffalo are common STEC reservoirs and can be colonized by a diversity of STEC serogroups possessing *stx* subtypes associated with severe disease in humans. STEC in buffalo can be transmitted to humans via meat, and unpasteurised milk and cheeses (Lorusso *et al.*, 2009; Vu-Khac and Cornick, 2008).

Currently, no specific STEC interventions are applied or suggested to reduce STEC in water buffalo during primary production through to product consumption. However, to control STEC contamination in raw buffalo milk intended for the production of mozzarella cheese, some limited studies demonstrated that heating curd during stretching produced a reduction of STEC (Trevisani, Mancusi and Valero, 2014). Other studies have shown that use of hot wash water reduced initial bacterial load on carcasses substantially and improved the microbiological quality of buffalo meat (Sachinda, Sakhare and Rao, 1998). Similarly, aerobic plate counts on water buffalo meat product was reduced by lactic acid spray (Manzoor, Jaspal and Yaqub, 2020).

> *GAP and GHP are suggested during water buffalo primary production and meat and milk processing; however, there was limited available evidence to support interventions specific for the control of STEC in water buffalo or water buffalo meat and dairy processing.*

6.3.5. Bison

Bison (*Bison bison*) are large ruminants that typically roam freely mainly in the United States of America and Canada. They are primarily fed on hay or grass and are slaughtered at about 18 months of age. In the United States of America, bison are considered an exotic species, so the federal inspection of the slaughter process is voluntary. There are some commercial feedlots for finishing bison in Canada, but little information exists on STEC carriage in these animals. Similar to other ruminants, bison can be colonized by STEC O157:H7 and non-O157 STEC (O121, O145) (Reinstein *et al.*, 2007; Magwedere *et al.*, 2013). In the United States of America, ground bison meat has been implicated in a multistate outbreak of STEC non-O157 (O103 and O121) (FDA, 2019).

> *GAP and GHP are suggested during bison primary production and meat processing; however, there was limited available evidence to support interventions specific for the control of STEC in bison or bison meat processing.*

6.3.6. Wild game

Wild game (deer, wild boar and hare) are reservoirs for STEC serotypes linked to human illness. Currently, no specific interventions against STEC have been validated for wild game small ruminants. However, it is thought that GHP and various processing/post-processing interventions that have been used to reduce

NTS *E. coli* and STEC levels on farms and along the cattle food chain are most likely to be effective against STEC that colonise wild game (Miko *et al.*, 2020).

Use of GHP is suggested during wild game meat processing; however, there was limited available evidence to support interventions specific for the control of STEC in wild game meat processing.

6.3.7. Swine

Several studies have demonstrated that domestic swine can carry and shed STEC and it may serve as vectors in human STEC outbreaks linked to fresh produce and other row crops (e.g. lettuce and spinach). Many studies worldwide have identified a low prevalence of STEC O157:H7 in swine, however results are conflicting (Tseng *et al.*, 2014; Magwedere *et al.*, 2013). In some studies, the prevalence of non-O157 STEC was high and included serogroups associated with severe disease in humans. However, many of these STEC carried the stx_{2e} subtype which may cause edema disease in swine, but so far, has not been associated with severe disease in humans. Pork products have occasionally been confirmed as vehicles of STEC transmission (Mughini-Gras *et al.*, 2018), but it remains unknown whether the STEC contamination on pork was natural or came from processing or via cross-contamination from other foods (Colello *et al.*, 2016).

In swine production, the entry of STEC into swine herds may be limited by appropriate biosecurity measures (e.g. good feed hygiene, keeping swine herds separate from wild and production animals). During processing and post-processing of pork, there are no practices specific to control STEC, but all of the GHP used in pork production are thought to be similarly effective against STEC as they are against *Salmonella* (Colello *et al.*, 2016).

General interventions throughout the pork production chain described in the FAO/WHO report; "Interventions for the control of non-typhoidal Salmonella spp. in beef and pork" (FAO and WHO, 2016) are also expected to be effective against STEC. However, there was limited available evidence to support interventions specific for the control of STEC in swine prmary production or pork processing.

References

Adamu, M. S., Ugochukwu, I. C. I., Idoko, S. I., Kwabugge, Y. A., Abubakar, N. S. & Ameh, J. A. 2018. Virulent gene profile and antibiotic susceptibility pattern of Shiga toxin-producing *Escherichia coli* (STEC) from cattle and camels in Maiduguri, North-Eastern Nigeria. *Tropical Animal Health Production*, 50: 1327–1341. doi: 10.1007/s11250-018-1565-z.

Bai, X., Zhao, A., Lan, R., Xin, Y., Xie, H., Meng, Q., *et al.* 2013. Shiga Toxin-Producing *Escherichia coli* in Yaks (*Bos grunniens*) from the Qinghai-Tibetan Plateau, China. *PLOS ONE*, 8(6): e65537. doi: 10.1371/journal.pone.0065537.

Bandyopadhyay, S., Biswas, T. K., Sasmal, D., Ghosh, M. K., Dutta, T. K., Das S. C., Bhattacharya, D., Bera, A. K., Bandyopadhyay, S. & De, S. 2009. Virulence gene and antibiotic resistance profile of Shiga-toxin-producing *Escherichia coli* prevalent in captive yaks (*Poephagus grunniens*). *Veterinary Microbiology*, 138: 403–404. doi: 10.1016/j.vetmic.2009.04.016.

Bandyopadhyay, S., Lodh, C., Rahaman, H., Bhattacharya, D., Bera, A. K., Ahmed, F. A., Mahanti, A., Samanta, I., Mondal, D. K., & Bandyopadhyay, S. 2012. Characterization of Shiga toxin producing (STEC) and enteropathogenic *Escherichia coli* (EPEC) in raw yak (*Poephagus grunniens*) milk and milk products. *Research in Veterinary Science*, 93: 604–610. doi: 10.1016/j.rvsc.2011.12.011.

Baschera, M., Cernela, N. & Stevens, M. J. A. 2019. Shiga toxin-producing *Escherichia coli* (STEC) isolated from fecal samples of African dromedary camels. *One Health*, 7: 100087. doi: 10.1016/j.onehlt.2019.100087.

Callaway, T., Edrington, T., Anderson, R., Genovese, K., Poole, T., Elder, R., Byrd, J., Bischoff, K. & Nisbet, D. 2003. *Escherichia coli* O157:H7 populations in sheep can be reduced by chlorate supplementation. *Journal of Food Protection*, 66: 194–199. doi: 10.4315/0362-028X-66.2.194.

Callaway, T., Edrington, T., Brabban, A., Anderson, R., Rossman, M., Engler, M., Carr, M., Genovese, K., Keen, J., Looper, M., Kutter, E. & Nisbet, D. 2008. Bacteriophage isolated from feedlot cattle can reduce *Escherichia coli* O157:H7 populations in ruminant gastrointestinal tracts. *Foodborne Pathogens and Disease*, 5: 183–191. doi: 10.1089/fpd.2007.0057.

Colello, R., Cáceres, M. E., Ruiz, M. J., Sanz, M., Etcheverría, A. I. & Padola, N. L. 2016. From farm to table: follow-up of Shiga toxin-producing *Escherichia coli* throughout the pork production chain in Argentina. *Frontiers in Microbiology*, 7: 93–99. doi: 10.3389/fmicb.2016.00093.

Dorsa, W. J., Cutter, C. N., Siragusa, G. R. & Koohmaraie, M. 1996. Microbial decontamination of beef and sheep carcasses by steam, hot water spray washes, and a steam-vacuum sanitizer. *Journal of Food Protection*, 59: 127–135. doi: 10.4315/0362-028X-59.2.127.

Edrington, T. S., Callaway, T. R., Bischoff, K. M., Genovese, K. J., Elder, R. O., Anderson, R. C. & Nisbet, D. J. 2003. Effect of feeding the ionophores monensin and laidlomycin propionate and the antimicrobial bambermycin to sheep experimentally infected with *E. coli* O157:H7 and *Salmonella* Typhimurium. *Journal of Animal Science*, 81: 553–560. doi: 10.2527/2003.812553x

EFSA (European Food Safety Authority). 2013. Scientific Opinion of the panel on Biological Hazards on the public health hazards to be covered by inspection of meat from sheep and goats. *EFSA Journal*, 11: 3265. doi: 10.2903/j.efsa.2013.3265

FAO & WHO. 2016. Interventions for the control of non-typhoidal *Salmonella* spp. in beef and pork: Meeting report and systematic review. *Microbiological Risk Assessment Series* No. 30. Rome.

FAO & WHO. 2019. Attributing illness caused by Shiga toxin-producing *Escherichia coli* (STEC) to specific foods: Report. *Microbiological Risk Assessment Series* No. 32. Rome.

FDA (United States of America Food and Drug Administration). 2019. Outbreak investigation of *E. coli*: Ground Bison (July 2019). In: *US FDA, Outbreaks of foodborne illness.* Cited 10 November 2021. www.fda.gov/food/outbreaks-foodborne-illness/outbreak-investigation-e-coli-ground-bison-july-2019

Furlan, J. P. R., Gallo, I. F. L., Pires de Campos, A. C. L., Navarro, A., Kobayashi, R. K. T., Nakazato, G. & Stehling, E. G. 2019. Characterization of non-O157 Shiga toxin-producing *Escherichia coli* (STEC) obtained from feces of sheep in Brazil. *World Journal of Microbiology and Biotechnology*, 35: 134. doi: 10.1007/s11274-019-2712-z

Gutta V. R, Kannan, G., Lee, G. H., Kouakou, B. & Getz, R. 2009. Influences of short-term pre-slaughter dietary manipulation in sheep and goats on pH and microbial loads of gastrointestinal tract. *Small Ruminant Research*, 81: 21–28. doi: 10.1016/j.smallrumres.2008.10.008

Hassan, A. A., Skjerve, E., Bergh, C. & Nesbakken, T. 2015. Microbial effect of steam vacuum pasteurisation implemented after slaughtering and dressing of sheep and lamb. *Meat Science*, 99: 32–37. doi: 10.1016/j.meatsci.2014.08.007

Jacob, M. E., Foster, D. M., Rogers, A. T., Balcomb, C. C., Shi, X. & Nagaraja, T. G. 2013. Evidence of non-O157 Shiga toxin–producing *Escherichia coli* in the feces of meat goats at a US slaughter plant. *Journal of Food Protection*, 76: 1626–1629. doi: 10.4315/0362-028X.JFP-13-064

Kudva, I. T., Hunt C. W., Williams, C. J., Nance, U. M. & Hovde, C. J. 1997. Evaluation of dietary influences on *Escherichia coli* O157: H7 shedding by sheep. *Applied and Environmental Microbiology*, 63: 3878–3886. doi: 10.1128/AEM.63.10.3878-3886.1997

Laaksonen, S., Oksanen, A., Julmi, J., Zweifel, C., Fredriksson-Ahomaa, M. & Stephan, R. 2017. Presence of foodborne pathogens, extended-spectrum β-lactamase -producing *Enterobacteriaceae*, and methicillin-resistant *Staphylococcus aureus* in slaughtered reindeer in northern Finland and Norway. *Acta Veterinaria Scandanavia*, 59: 2. doi: 10.1186/s13028-016-0272-x

Li, Q., Sherwood, J. S. & Logue, C. M. 2004. The prevalence of *Listeria, Salmonella, Escherichia coli* and *E. coli* O157:H7 on bison carcasses during processing. *Food Microbiology*, 21: 791–799. doi: 10.1016/j.fm.2003.12.006

Lorusso, V., Dambrosio, A., Quaglia, N. C., Parisi, A., *et al.* 2009. Verocytotoxin-Pproducing *Escherichia coli* O26 in raw water buffalo (*Bubalus bubalis*) milk products in Italy. *Journal of Food Protection*, 72: 1705–1708. doi: 10.4315/0362-028x-72.8

Magwedere, K., Dang, H. A., Mills, E. W., Cutter, C. N., Roberts, E. L. & Debroy, C. 2013. Incidence of Shiga toxin–producing *Escherichia coli* strains in beef, pork, chicken, deer, boar, bison, and rabbit retail meat. Journal of Veterinary Diagnostic Investigation, 25: 254–8. doi: 10.1177/1040638713477407

Manzoor, A., Jaspal, M. H. & Yaqub, T. 2020. Effect of lactic acid spray on microbial and quality parameters of buffalo meat. *Meat Science*, 59: 107923. doi: 10.1016/j.meatsci.2019.107923

Miko, A., Pries, K., Haby, S., Steege, K., Albrecht, N., Krause, G. & Beutin L. 2009. Assessment of Shiga toxin-producing *Escherichia coli* isolates from wildlife meat as potential pathogens for humans. *Applied and Environmental Microbiology*, 75: 6462–6470. doi: 10.1128/aem.00904-09

Mughini-Gras, L., van Pelt, W., van der Voort, M., Heck, M., Friesema, I. & Franz, E. 2018. Attribution of human infections with Shiga toxin-producing *Escherichia coli* (STEC) to livestock sources and identification of source-specific risk factors, The Netherlands (2010-2014). *Zoonoses and Public Health*, 65: e8–e22. doi: 10.1111/zph.12403

Pointon, A., Kiermeir, A. & Fegan, N. 2012. Review of the impact of preslaughter feed curfews of cattle, sheep, and goats on food safety and carcass hygiene in Australia. *Food Control*, 26: 313–321. doi: 10.1016/j.foodcont.2012.01.034

Raya, R. R., Oot, R. A., Moore-Maley, B., Wieland, S., Callaway, T. R., Kutter, E. M. & Brabban, A. D. 2011. Naturally resident and exogenously applied T4-like and T5-like bacteriophages can reduce *Escherichia coli* O157:H7 levels in sheep guts. *Bacteriophage*, 1: 15–24. doi: 10.4161/bact.1.1.14175

Rehman, M. U., Zhang, H., Wang, Y., Mehmood, K., Huang, S., Iqbal, M. K. & Li, J. 2017. Experimental mouse lethality of *Escherichia coli* strains isolated from free ranging Tibetan yaks. *Microbial Pathogens*, 10: 15–19. doi: 10.1016/j.micpath.2017.05.020

Reinstein, S., Fox, J. T., Shi, X., Alam, M. J. & Nagaraja, T. G. 2007. Prevalence of *Escherichia coli* O157:H7 in the American Bison (*Bison bison*). *Journal of Food Protection*, 70: 2555–2560. doi: 10.4315/0362-028x-70.11.2555

Rigobelo, E. E. C., Karapetkov, N., Maestai, S. A., Avila, F. A. & McIntosh, D. 2015. Use of probiotics to reduce faecal shedding of Shiga toxin-producing *Escherichia coli* in sheep. *Beneficial Microbes*, 6: 53–60. doi: 10.3920/BM2013.0094

Sachindra, N. M., Sakhare, P. Z. & Rao D. N. 1998. Reduction in microbial load on buffalo meat by hot water dip treatment. *Meat Science*, 48: 149–157. doi: 10.1016/s0309-1740(97)00085-5

Salehi, T. Z., Tonelli, A., Mazza, A., Staji, H., Badagliacca, P., Tamai, I. A., Jamshidi, R., Harel, J., Rossella, L. & Masson, L. 2012. Genetic characterization of *Escherichia coli* O157:H7 strains isolated from the one-humped camel (*Camelus dromedarius*) by using microarray DNA technology. *Molecular Biotechnology*, 51: 283–288. doi: 10.1007/s12033-011-9466-7

Tabatabaei, S., Salehi, T. Z., Badouei, M. A., Tamai, I. A., Akbarinejad, V., Kazempoor, R. & Shojaei, M. 2013. Prevalence of Shiga toxin-producing and enteropathogenic *Escherichia coli* in slaughtered camels in Iran. *Small Ruminant Research*, 113: 297–300. doi: 10.1016/j.smallrumres.2012.12.011

Thomas, C. L., Stelzleni, A. M., Rincon, A. G., Kumar, S., Rigdon, M., McKee, R. W. & Thippareddi, H. 2019. Validation of antimicrobial interventions for reducing Shiga toxin–producing *Escherichia coli* surrogate populations during goat slaughter and carcass chilling. *Journal of Food Protection*, 82: 364–370. doi: 10.4315/0362-028x.jfp-18-298

Trevisani, M., Mancusi, R. & Valero, A. 2014. Thermal inactivation kinetics of Shiga toxin-producing *Escherichia coli* in buffalo Mozzarella curd. *Journal of Dairy Science*, 97: 642–650. doi: 10.3168/jds.2013-7150

Tseng, M., Fratamico, P. M., Manning, S. D. & Funk, J. A. 2014. Shiga toxin-producing *Escherichia coli* in swine: the public health perspective. *Animal Health Research Reviews*, 15: 63–75. doi: 10.1017/S1466252313000170

Vu-Khac, H. & Cornick, N. A. 2008. Prevalence and genetic profiles of Shiga toxin-producing *Escherichia coli* strains isolated from buffaloes, cattle, and goats in central Vietnam. *Veterinary Microbiology*, 126: 356–363. doi: 10.1016/j.vetmic.2007.07.023

Wang L, Qu, K., Li, X., Cao, Z., Wang, X, Li, Z., Song, Y. & Xu, Y. 2017. Use of Bacteriophages to control *Escherichia coli* O157:H7 in domestic ruminants, meat products, and fruits and vegetables. *Foodborne Pathogens and Disease*, 14: 483–493. doi: 10.1089/fpd.2016.2266

Yekta, M. A., Cox, E., Goddeeris, B. M. & Vanrompay, D. 2011. Reduction of *Escherichia coli* O157:H7 excretion in sheep by oral lactoferrin administration. *Veterinary Microbiology*, 150: 373–378. doi: 10.1016/j.vetmic.2011.02.052

Zhang, X., Yu, Z., Zhang, S. & He, K. 2014. Immunization with H7-HCP-tir-intimin significantly reduces colonization and shedding of *Escherichia coli* O157:H7 in goats. *PLOS ONE*, 9: e91632. doi: 10.1371/journal.pone.0091632

Zweifel, C., Fierz, L., Cernela, N., Laaksonen, S., Fredriksson-Ahomaa, M. & Stephan, R. 2017. Characteristics of shiga toxin-producing *Escherichia coli* O157 in slaughtered reindeer from northern Finland. *Journal of Food Protection*, 80: 454–458. doi: 10.4315/0362-028X.JFP-16-457

7

Laboratory testing

7.1. LABORATORY TESTING FOR STEC DURING PRIMARY PROCESSING

The expert committee concluded that the implementation of a monitoring plan at the cattle farm level to measure the impact on STEC prevalence in raw beef is impractical.

This conclusion is based on several factors, including:

- Excretion of STEC in faeces is highly intermittent, so the value of predicting risk from the collection of a single faecal sample is questionable.
- Super shedders are likely responsible for the majority of STEC transmission within the herd, but currently, there is no reliable method for the rapid detection of super shedder cattle.
- The degree of STEC excretion varies substantially among individual animals, hence, in order to gain a true measurement of prevalence, it will be necessary to sample single animals individually as opposed to collecting composited faecal samples from the pen.
- Detecting all the possible STEC serogroups simultaneously in cattle has proven to be extremely challenging. At present, there is no test that can detect all the STEC serogroups other than the "top seven" serotypes that may be present in a faecal sample collected from an animal.
- Some available STEC detection methods are relatively sophisticated and not easily implemented on a farm. Consequently, a significant investment for a centralized laboratory infrastructure and trained personnel will be required to support on farm monitoring.

- Testing a single faecal sample for STEC is inadequate and testing multiple faecal samples from a single animal would be extremely costly. The need to repeatedly restrain the animal to collect multiple faecal samples would raise animal welfare concerns.
- If an animal was found to be STEC positive, it would be difficult to manage it within the production chain, as delaying slaughter would substantially increase production costs and will likely require isolating the STEC positive animal by housing it away from the main herd to reduce the risk of STEC transmission. Such a practice could become costly and adversely impact animal health and productivity and meat quality.
- Even if the animals were found to be STEC negative at the time of sampling, it would not be advisable to reduce processing and post-processing intervention strategies, as many of these practices are broad-based and intended to prevent a variety of pathogens from entering the food chain.

7.2. LABORATORY TESTING FOR STEC DETECTION ACROSS THE BEEF PROCESSING CHAIN

It is clear that in general, the occurrence of STEC in meats is lower for intact meat products than in trim or ground/minced beef (Kintz *et al.*, 2017; Devleesschauwer *et al.*, 2019). Intact meats originate from a single animal, whereas ground/minced beef is composed of meat pieces from many animals, so one piece of meat contaminated with STEC can spread the contamination to the entire batch. However, the overall occurrence of STEC in these products can vary considerably due to the differences in primary, processing and post-processing conditions and the interventions applied. As a result, it is difficult to set STEC criteria for laboratory analysis for all possible conditions.

Laboratory testing to monitor for STEC at beef processing and post-processing stages depends on a country's regulation or export requirements and the testing methods used may also vary. For example, there are methods that use a stepwise approach to first screen for Shiga toxin genes (*stx*) and the *E. coli* attaching and effacing gene (*eae*), followed by testing for major O serogroup–specific genes (e.g. USDA, 2021 ISO/TS 13136-2012). A separate method is used to only test for STEC O157:H7 (e.g. ISO 16654:2001). According to the JEMRA report on STEC risk characterization, testing for STEC virulence genes (*stx*, *eae* and *aggR*), regardless of the serogroup, was recommended as the serotype does not necessarily predict the virulence profile (FAO and WHO, 2018).

For reliable STEC detection in food, methods accredited or validated by independent or official organizations are required. The many different STEC detection methods include microbiological culture enrichment, selective/chromogenic media, immunological Enzyme linked immunosorbent assay (ELISA), Immunomagnetic separation (IMS), Lateral flow, Latex agglutination, Microplate enzyme immunoassay (EIA), Optical immunoassay, Reverse passive latex agglutination (RPLA), and molecular (polymerase chain reaction [PCR], Real-time PCR, Loop-mediated isothermal amplification PCR [LAMP], digital droplet PCR [ddPCR]) methods. Many, but not all of these methods have been validated.

Culture enrichment is a critical step in most STEC detection methods and an indispensable requirement in order to detect the low numbers of STEC that maybe present amidst the high-level of background cells in meats and other foods. Improved selective media formulations that facilitate STEC enrichment can decrease the sample-to-result time and improve sensitivity. Improved enrichment procedures coupled with screening tools (e.g. Shiga toxin EIA, PCR, LAMP) can be used to obtain definitive negative or presumptive positive results more rapidly. At present, more culture media to enhance the enrichment of STEC O157:H7 have been developed (Bai and Xiong, 2019) than for non-O157 STEC strains (Conrad *et al.*, 2016; Parsons *et al.*, 2016). Further work is required to develop and validate enrichment procedures for all STEC serogroups of health concern (Brusa, Piñeyro and Galli, 2016; Castro *et al.*, 2017; NACMCF, 2019).

The use of assays, such as PCR, that target STEC virulence gene in foods is recommended to, as they demonstrate both high sensitivity and specificity, and are also fast, low cost and commercially available. Currently, real-time PCR, LAMP and dd-PCR that target *stx* and/or *eae* genes are commercially available and some have been validated. Most *stx* PCR assays can detect *stx* subtypes often associated with severe disease, but will miss genetically more distant *stx* subtypes, some of which have been associated with severe human disease (Paton and Paton, 1998; Feng *et al.*, 2011; Reischl *et al.*, 2002; Beutin, Jahn and Fach, 2009; Scheutz *et al.*, 2012). It is also important to realize that bacteria other than STEC may harbour some of these virulence genes. Furthermore, the mere presence of a virulence gene may not be reflective of health risk due to differential or lack of *stx* gene expression. Subtyping the *stx* gene variants is also important in order to discriminate subtypes most often associated with human disease from those that may not cause human infections (Scheutz *et al.*, 2012; Staten Serum Institute, 2014; FAO and WHO, 2018).

Isolation of the STEC organism from presumptive positive samples is a requirement for confirmation and this is often performed with traditional culture-based methods. Not all presumptive positive samples can be confirmed, but the use of IMS techniques have greatly improved the isolation of STEC from enrichment cultures. However, IMS is currently available only for the *eae* positive STEC serogroups commonly associated with human diseases. Furthermore, IMS are very serogroup-specific, but some IMS assays lack target specificity, and may capture strains from more than one serogroup or fail to identify other STEC serogroups (Kraft *et al.*, 2017). Immunoblot and colony hybridization protocols have recently been developed that can enhance and improve STEC isolation from a variety of food matrices.

Testing food for STEC as part of monitoring programmes is of limited use due to the low level of STEC found in foods. Hence, the quantitative detection of NTS *E. coli* (ISO 16649-2) as a process hygiene indicator, is proposed as an alternative approach to monitor hygiene during processing and post-processing. NTS *E. coli* counts can vary considerably from sample to sample or plant to plant, so internal company standards or specific criteria recommendations are not possible for all meat products. The criteria for decision making based on hygiene indicator levels can vary depending on the pre-defined limit set and the sampling plans implemented, but they could still be useful for trend analysis with regard to STEC surveys and for STEC baseline studies (ICMSF, 2011).

Several novel technologies (e.g. LAMP, recombinase polymerase amplification [RPA], PCR-mass spectrometry, whole genome sequencing [WGS]), are being explored for STEC detection (Pierce *et al.*, 2012). WGS for STEC detection and typing uses different high throughput sequencing platforms (Illumina, Oxford Nanopore technology, Ion Torrent and Pacific Biosciences [PacBio]) but is becoming increasingly common and accepted (Worley *et al.*, 2017; Allard *et al.*, 2018; Wilson *et al.*, 2018; Mylius *et al.*, 2018; González-Escalona *et al.*, 2019; Franz *et al.*, 2014). However, issues associated with the implementation and long-term sustainability of WGS, such as cost, equipment, maintenance, supplies, IT, training, are limitations for many countries. Further work is needed to develop rapid and standardized analysis protocol for the various WGS methodologies so that the data are comparable and these methods also need to be validated. Additional experimental data are also needed on these (and other) novel technologies to support their potential application as methods for the identification and characterization of STEC. Biosensors (Subramanian *et al.*, 2012; Pandey *et al.*, 2017), mass spectrometry (MALDI-TOF MS; Pierce *et al.*, 2012), and nanotechnology (Jyoti *et al.*, 2010) are examples of emerging future technologies that may be explored for detection and characterization of STEC from foods.

7.3. LABORATORY TESTING FOR STEC DETECTION ACROSS THE DAIRY PROCESSING CHAIN

Sampling and analysis of raw milk and raw milk products are important steps within the verification plans to confirm that the practices and procedures implemented in the food safety program have been met.

7.3.1. Raw milk

Although STEC has been isolated from raw milk, STEC testing of milk is uncommon and most sampling and testing protocols target indicator organisms such as *E. coli*. Whereas the presence or concentration of NTS *E. coli* or other indicator organisms in raw milk is not indicative of the presence of STEC, they remain useful hygienic markers of the quality of raw milk (Metz, Sheehan and Feng, 2019). Sampling and testing plans for raw milk are highly dependent on consumption practices, the scale of production, local vs regional regulations, and as such, are highly variable between countries and across regions.

At the farm level, given the impracticality of sampling milk from each producer or from every animal daily, it is essential to use a well-designed sampling program that can reduce cost and time but also provide sufficient data to adequately assess the hygiene of milk in bulk tanks. Depending on the size of the milk collection operation (small-, medium- or large-scale) different sampling approaches are available, including periodic, random, composite and universal sampling. Sampling may be done once a week or once every two weeks or done periodically at irregular intervals. Raw milk can also be sampled and tested on a random basis (random sampling), or samples may be pooled together over a period of time (composite samples) and tested.

For large-scale operations, a universal sampling system is used by bulk milk haulers every time raw milk is picked up at the farm and collected. Aliquots of the universal milk sample are sent to an approved laboratory for analyses (FDA/USPHS, 2017).

Entry of STEC via contaminated raw milk into dairy food processing plants can lead to persistence of pathogens in biofilms, exposure of consumers to STEC in unpasteurized dairy products as well as subsequent contamination of other processed milk products (Oliver, Jayarao and Almeida, 2005). Thus, samples from valves, equipment, filters and environmental sources are routinely collected at plants for microbiological analysis. The universal sampling system permits the competent authority, at any given time and without notification to the industry, to analyze samples collected by the bulk milk hauler/sampler and/or industry plant sampler at the farm and milk processing plant, respectively (Oliver, Jayarao and Almeida, 2005).

Although most testing has focused on sampling of bulk tank milk, studies have also assessed the utility of sampling milk filters. Jaakkonen *et al.* (2019) conducted a 1-year longitudinal study on the presence of STEC and *Campylobacter jejuni* on Finnish dairy farms and in raw milk. STEC O157:H7 was isolated from 17 percent of the cattle, but from only 2 percent of milk filters and not from any samples of raw milk. However, the *stx* gene was detected at a higher frequency from milk filters (37 percent) than in raw milk (7 percent), suggesting that the filters may be a more effective sampling point than raw milk. Artursson *et al.* (2018) also observed that in-line milk filters were a better sampling point for the presence of pathogens in general than sampling the milk in the bulk tanks.

Some of the logistical limitations to bulk tank and plant sampling include: the duration of agitation to ensure adequate mixing, the need to use aseptic sampling techniques and the need to use disinfected sampling equipment and containers. Additional factors to consider include, sample preservation methods, storage temperatures (raw milk should be cooled to 7 °C or less within 2 h after the completion of milking) and the collection of appropriate information, which are necessary to ensure proper sampling, storage, transportation and identification/tracing of the samples.

7.3.2. Raw milk cheeses

The microbiological safety of raw milk cheese is managed by the effective implementation of control measures that have been validated, where appropriate, to minimize contamination from the milking process through to the maturation of the cheese. This preventative approach is more effective than relying solely on microbiological testing of individual final product lots for market acceptance. However, setting microbiological criteria may be appropriate for verifying that food safety control systems were implemented correctly (FAO and WHO, 2013). The established microbiological criterion should be based on risk assessment taking into considerations such factors as epidemiological evidence. The criterion should be meaningful for consumer protection, so if a criteria is not met, the cheese in question may represent a significant public health risk.

Consideration should also be given to the behaviour of STEC during the cheese-making and maturation process, because STEC prevalence changes over the course of manufacturing, distribution, storage, marketing and preparation. There are strain to strain variations, and some pathogenic strains of STEC may be more acid tolerant than NTS *E. coli* and persist when the indicator has died off. So, any microbiological criterion set must be established at a specified point in the food chain (FAO and WHO, 2013). Consideration needs to be given to sampling plans that can effectively remove highly contaminated lots and result in continuous improvement without completely disrupting the food supply (ICMSF, 2001). Perrin *et al.* (2015) conducted a risk assessment of soft cheeses made from raw milk and considered the effect of applying various microbiological criteria at the end of ripening. They found that various criteria (e.g. differing in terms of sample size, the number of samples that may yield a value larger than the microbiological limit, and the methods for STEC detection) could reduce the risk of STEC induced hemolytic uremic syndrome (HUS) in human patients by 25 to 89 percent. Increasing the sample size of the end-product for analysis from 25 g to 100 g for STEC testing was also predicted to reduce HUS risk (ANSES opinion, 2018). Risk managers must balance risk reduction with economics so that the products remain available for sale at a reasonable cost.

Consequently, microbiological criteria should be considered to be part of a food safety control system and should also include ongoing monitoring of the system. Although there is strain to strain variations, some pathogenic strains of STEC may be more acid tolerant than NTS *E. coli* and persist when the indicator has died off.

There are general recommendations in microbiological sampling programs to assure the hygiene and safety of cheeses, but none are specific for STEC. In cheeses made from pasteurized milk, the International Commission on Microbiological Specifications for Foods (ICMSF) (2011) recommends the use of *E. coli* limits that are established under a 3-class sampling plan, where n (sample number) = 5, c (allowable number of samples between m and M) = 3, m (acceptable level) = 10 and M (unacceptable level) = 10^2. Raw milk cheeses are tested for *Staphylococcus aureus* only, which is consistent with the European Union recommended sampling criteria. The ICMSF provides a suggested sampling plan for *Salmonella* in raw milk cheese (medium or low importance) where n = 5 (25 g samples), c = 0, m = nd (not detected), but makes no STEC recommendations.

As STEC prevalence in raw milk and raw milk cheeses is low, STEC testing in these products is uncommon and challenging, and most sampling and testing protocols target indicator organisms such as *E. coli*. It is notable that for the European Union microbiological criteria for cheese, no limits were established for *E. coli* in raw milk cheese. It is regarded that *E. coli* does not offer a meaningful hygienic index in raw milk cheese as its presence is expected, consistent with guidance from ICMSF. However, the Health Protection Agency (the United Kingdom) recommends that raw milk cheese be tested routinely for *E. coli*, and if detected, the source of contamination investigated, particularly if an upward trend is noted since STEC may also be present (Donnelly, 2018). Although the presence of NTS *E. coli* or other indicator organisms in raw milk does not indicate the presence of STEC, they remain useful hygienic markers of the quality of raw milk and raw milk cheeses and many other countries in the world have also established NTS *E. coli* limits to monitor the sanitary quality of raw milk cheeses (Metz *et al.*, 2019).

Monitoring of raw milk cheese quality is not a true intervention step, however, it can contribute to the safety of raw milk cheeses. Systematic selection of quality raw milk with < 50 CFU/mL of *E. coli* for use in the manufacturing of uncooked, pressed cheese was predicted to reduce the risk of HUS from STEC (ANSES opinion, 2018). Depending on the cheese types and the technology used, quality guidelines for the manufacturer of raw milk cheeses, which though may not fully eliminate STEC from the product, may include STEC relevant microbial criteria that contribute to the safety of raw milk cheese products. However, setting such criteria is complex, as there is a large variety of cheese types and wide diversity in behaviour among STEC strains. As a result, challenge tests with various STEC strains in various raw milk cheese types need to be performed to investigate the survival of these pathogens under those specific manufacturing conditions. In the absence of official criteria for STEC or *E. coli* (in some countries) in raw milk cheeses, small and medium enterprises that make a specific cheese type or a group of similar products, and where real-time monitoring of *E. coli* in raw milk is not feasible, a specific limit for *E. coli* in young 24-hour old cheese (after pressing, before brining) should be defined. By using quality raw milk, in combination with proper manufacturing and ageing for at least 60 days, and assuming a monthly decrease of 1 $\log_{10}$ CFU/g, it is possible to make ready-to-eat raw milk cheeses that are absent of *E. coli* or only contain low numbers (Metz, Sheehan and Feng, 2019).

References

Allard, M.W., Bell, R., Ferreira, C.M., Gonzalez-Escalona, N., Hoffmann M. & Muruvanda, T. 2018. Genomics of foodborne pathogens for microbial food safety. *Current Opinions in Biotechnology*, 49: 224–229. doi: 10.1016/j.copbio.2017.11.002

ANSES (Agence Nationale Sécurité Sanitaire Alimentaire Nationale). 2018. Avis de l'Anses relatif au protocole de reprise de la commercialisation de reblochons proposé par l'entreprise Chabert. 2018. www.anses.fr/fr/system/files/BIORISK20 18SA0164.pdf

Artursson, K., Schelin, J., Thisted Lambertz, S., Hansson, I. & Olsson Engvall, E. 2018. Foodborne pathogens in unpasteurized milk in Sweden. *International Journal of Food Microbiology*, 284: 120–127. doi: 10.1016/j.ijfoodmicro.2018.05.015

Bai, X. & Xiong, Y. 2019. Part II: Foodborne bacterial pathogens. *Escherichia coli* O157:H7. In Hu, L, ed. *Food safety: Rapid detection and effective prevention of foodborne hazards.* Apple Academic Press. ISBN: 9781774630686.

Beutin, L., Jahn, S. & Fach, P. 2009. Evaluation of the 'GeneDisc' real-time PCR system for detection of enterohaemorrhagic *Escherichia coli* (EHEC) O26, O103, O111, O145 and O157 strains according to their virulence markers and their O- and H-antigen-associated genes. *Journal of Applied Microbiology*, 106: 1122–1132. doi: 10.1111/j.1365-2672.2008.04076.x

Brusa, V., Piñeyro, P. E. & Galli, L. 2016. Isolation of Shiga toxin-producing *Escherichia coli* from ground beef using multiple combinations of enrichment broths and selective agars. *Foodborne Pathogens and Disease*, 13: 163–170. doi: 10.1089/fpd.2015.2034

Castro, V.S., Carvalho, R. C. T., Conte-Junior, C. A. & Figuiredo, E. E. S. 2017. Shiga-toxin producing *Escherichia coli*: pathogenicity, supershedding, diagnostic methods, occurrence, and foodborne outbreaks. *Comprehensive Reviews in Food Science and Food Safety*, 16: 1269–1280. doi: 10.1111/1541-4337.12302

Conrad, C. C., Stanford, K., McAllister, T. A., Thomas, J. & Reuter, T. 2016. Competition during enrichment of pathogenic *Escherichia coli* may result in culture bias. *FACETS*, 1: 114–126. doi: 10.1139/facets-2016-0007

Devleesschauwer, B., Pires, S. M., Young, I., Gill, A. & Majowicz, S. E. 2019. Associating sporadic, foodborne illness caused by Shiga toxin-producing *Escherichia coli* with specific foods: a systematic review and meta-analysis of case-control studies. *Epidemiology and Infection*, 147: e235, 1–16. doi: 10.1017/S0950268819001183

Donnelly, C. 2018. Review of controls for pathogen risks in Scottish artisan cheeses made from unpasteurised milk. www.foodstandards.gov.scot/downloads/FSS_2017_015_-_Control_of_pathogens_in_unpasteurised_milk_cheese_-_Final_report_v_4.7_-_20th_November_2018_.pdf

FAO & WHO. 2013. Codex Alimentarius. Principles and guidelines for the establishment and application of microbiological criteria related to foods (CAC/GL 21-1997). Rome, FAO.

FAO & WHO. 2018. Shiga toxin-producing *Escherichia coli* (STEC) and food: attribution, characterization, and monitoring. *Microbiological Risk Assessment Series* No. 31. Rome. FAO.

FDA & USPHS (The United States Food and Drug Administration & Public Health Service). 2017. Grade "A" pasteurized milk ordinance. Public Health Service/Food and Drug Administration. 2019 Revision. Cited on 20 August 2022. www.fda.gov/media/140394/download

Franz, E., Delaquis, P., Morabito, S., Beutin, L., Gobius, K. & Rasko, D. A. 2014. Exploiting the explosion of information associated with whole genome sequencing to tackle Shiga toxin-producing *Escherichia coli* (STEC) in global food production systems. *International Journal of Food Microbiology*, 187: 57–72. doi: 10.1016/j.ijfoodmicro.2014.07.002

Feng, P. C. H, Jinneman, K., Scheutz, F. & Monday, S. R. 2011. Specificity of PCR and serological assays in the detection of *Escherichia coli* Shiga toxin subtypes. *Applied and Environmental Microbiology*, 77: 6699–6702. doi: 10.1128/AEM.00370-11

González-Escalona, N., Allard, M. A., Brown, E. W., Sharma, S. & Hoffmann, M. 2019. Nanopore sequencing for fast determination of plasmids, phages, virulence markers, and antimicrobial resistance genes in Shiga toxin-producing *Escherichia coli*. *PLOS ONE*, 14: e0220494. doi: 10.1371/journal.pone.0220494

ICMSF (International Commission on Microbiological Specifications for Foods). 2001. Microorganisms in Foods 7: Microbiological testing in food safety management. NY: Kluwer Academic / Plen Publishers.

ICMSF. 2011. Microorganisms in Foods 8: Use of data for assessing process control and product acceptance. New York, NY: Springer.

ISO/TS 13136. 2012. Microbiology of food and animal feed - real-time polymerase chain reaction (PCR)-based method for the detection of food-borne pathogens - horizontal method for the detection of Shiga toxin-producing *Escherichia coli* (STEC) and the determination of O157, O111, O26, O103 and O145 serogroups. International Organization for Standardization, Geneva, Switzerland.

ISO 16649-2:2001(E). 2012. Microbiology of food and animal feeding stuffs–Horizontal method for the enumeration of beta -glucuronidase -positive *E. coli*–Part 2: Colony count technique at 44°C using 5-bromo-4-chloro-3-indolyl beta-D-glucuronide. International Organization for Standardization, Geneva, Switzerland.

ISO 16654:2001. 2018. Microbiology of food and animal feeding stuffs — Horizontal method for the detection of *Escherichia coli* O157, International Organization for Standardization, Geneva, Switzerland.

Jaakkonen, A., Castro, H., Hallanvuo, S., Ranta, J., Rossi, M., Isidro, J., Lindström, M. & Hakkinen, M. 2019. Longitudinal study of Shiga toxin-producing *Escherichia coli* and *Campylobacter jejuni* on Finnish dairy farms and in raw milk. *Applied and Environmental Microbiology*, 85: e02910–18. doi: 10.1128/AEM.02910-18

Jyoti, A., Pandey, P., Singh, S. P., Jain, S. K. & Shanker, R. 2010. Colorimetric detection of nucleic acid signature of Shiga-toxin producing *Escherichia coli* using gold nanoparticles. *Journal of Nanoscience and Nanotechnology*, 10: 4154–4158. doi: 10.1166/jnn.2010.2649

Kintz, E., Brainard, J., Hooper, L. & Hunter, P. 2017. Transmission pathways for sporadic Shiga-toxin producing *E. coli* infections: A systematic review and meta-analysis. *International Journal of Hygiene and Environmental Health*, 220: 5767. doi: 10.1016/j.ijheh.2016.10.011

Kraft, A. L., Lacher, D. W., Shelver, W. L., Sherwood, J. S. & Bergholz, T. M. 2017. Comparison of immunomagnetic separation beads for detection of six non-O157 Shiga toxin-producing *Escherichia coli* serogroups in different matrices. *Letters in Applied Microbiology*, 65(3): 2013–219. doi: 10.1111/lam.12771

Metz, M., Sheehan, J. & Feng, P. 2019. Use of indicator bacteria for monitoring insanitation in raw milk cheeses – a Review. *Food Microbiology*, 85: 103283. doi: 10.1016/j.fm.2019.103283

Mylius, M., Dreesman, J., Pulz, M., Pallasch, G., Beyrer, K., Claußen, K., Allerberger, F., Fruth, A., Lang, C., Prager, R., Flieger, A., Schlager, S., Kalhöfer, D. & Mertens, E. 2018. Shiga toxin-producing *Escherichia coli* O103:H2 outbreak in Germany after school trip to Austria due to raw cow milk, 2017 - The important role of international collaboration for outbreak investigations. *International Journal of Medical Microbiology*, 308: 539–544. doi: 10.1016/j.ijmm.2018.05.005

National Advisory Committee on Microbiological Criteria for Foods. 2019. Response to questions posed by the Food and Drug Administration regarding virulence factors and attributes that define foodborne Shiga toxin-producing *Escherichia coli* (STEC) as severe human pathogens. *Journal of Food Protection*, 82: 724–767. doi: 10.4315/0362-028X.JFP-18-479

Oliver, S. P., Jayarao, B. M., & Almeida, R. A. 2005. Foodborne pathogens in milk and the dairy farm environment: food safety and public health implications. *Foodborne Pathogens and Disease*, 2(2), 115–129. doi: 10.1089/fpd.2005.2.115

Pandey, A., Gurbuz, Y., Ozguz, V., Niazi, J. H. & Qureshi, A. 2017. Graphene-interfaced electrical biosensor for label-free and sensitive detection of foodborne pathogenic *E. coli* O157:H7. *Biosensors and Bioelectronics*, 91: 225–231. doi: 10.1016/j.bios.2016.12.041

Parsons, B. D., Zelyas, N., Berenger, B. M. & Chui, L. 2016. Detection, characterization, and typing of Shiga Toxin-producing *Escherichia coli*. *Frontiers in Microbiology*, 7: 478. doi: 10.3389/fmicb.2016.00478

Paton, A. W. & Paton, J. C. 1998. Detection and characterization of Shiga toxin *Escherichia coli* by using multiplex PCR assays for *stx*1, *stx*2m, eaeA, enterohemorrhagic *E. coli* hylA, rfb O111, and rfb O157. *Journal of Clinical Microbiology*, 36(2): 598–602. doi: 10.1128/JCM.36.2.598-602.1998

Perrin, F., Tenenhaus-Aziza, F., Michel, V., Miszczycha, S., Bel, N. & Sanaa, M. 2015. Quantitative risk assessment of haemolytic and uremic syndrome linked to O157:H7 and non-O157:H7 Shiga-toxin-producing *Escherichia coli* strains in raw milk soft cheeses. *Risk Analysis*, 35: 109–128. doi: 10.1111/risa.12267

Pierce, S. E., Bell, R. L., Hellberg, R. S., Cheng, C. M., Chen, K. S., Williams-Hill, D. M., Martin, W. B. & Allard M. W. 2012. Detection and identification of *Salmonella enterica*, *Escherichia coli*, and *Shigella* spp. via PCR-electrospray ionization mass. *Applied and Environmental Microbiology*, 78: 8403–8411. doi: 10.1128/AEM.02272-12

Reischl, U., Youssef, M. T., Kilwinski, J., Lehn, N., Zhang, W. L., Karch, H., & Strockbine, N. A. 2002. Real-time fluorescence PCR assays for detection and characterization of Shiga toxin, intimin, and enterohemolysin genes from Shiga toxin-producing *Escherichia coli*. *Journal of Clinical Microbiology*, 40: 2555–2565. doi: 10.1128/jcm.40.7.2555–2565.2002

Scheutz, F., Teel, L. D., Beutin, L., Piérard, D., Buvens, G., Karch, H., Mellmann, A., Caprioli, A., Tozzoli, R., Morabito, S., Strockbine, N. A., Melton-Celsa, A. R., Sanchez, M., Persson, S. & O'Brien, A. D. 2012. Multicenter evaluation of a sequence-based protocol for subtyping Shiga toxins and standardizing Stx nomenclature. *Journal of Clinical Microbiology*, 50: 2951–2963. doi: 10.1128/JCM.00860-12

Statens Serum Institute. 2014. Identification of three *vtx1* and seven *vtx2* subtypes of Shiga toxin encoding genes of *Escherichia coli* by conventional PCR amplification. Cited 20 August 2022. www.ssi.dk/English/HealthdataandICT/National%20Reference%20Laboratories/Bacteria/~/media/2FCF8E537BF448F69C7FD63D6F953524

Subramanian, S., Aschenbach, K. H., Evangelista, J. P., Najjar, M. P., Song, W. & Gomez R. D. 2012. Rapid, sensitive and label-free detection of Shiga-toxin producing *Escherichia coli* O157 using carbon nanotube biosensors. *Biosensors and Bioelectronics*, 32: 69–75. doi: 10.1016/j.bios.2011.11.040

USDA (United States of America Department of Agriculture). 2021. Detection, isolation and identification of top seven Shiga toxin-producing *Escherichia coli* (STECs) from meat products and carcass and environmental sponges. Microbiological Laboratory Guidebook 5C.01. www.fsis.usda.gov/sites/default/files/media_file/2021-03/mlg-5.pdf

Wilson, D., Dolan, G., Aird, H., Sorrell, S., Dallman, T.J., Jenkins, C., Robertson, L. & Gorton, R. 2018. Farm-to-fork investigation of an outbreak of Shiga toxin-producing *Escherichia coli* O157. *Microbial Genomics*, 4: e000160. doi: 10.1099/mgen.0.000160

Worley, J. N., Flores, K. A., Yang, X., Chase, J. A., Cao, G., Tang, S., Meng, J. & Atwill, E. R. 2017. Prevalence and genomic characterization of *Escherichia coli* O157:H7 in cow-calf herds throughout California. *Applied and Environmental Microbiology*, 83: e00734–17. doi: 10.1128/AEM.00734-17

8

Conclusions

The expert committee performed a review of accessible scientific evidence on the efficacy and utility of physical, chemical and biological control measures effective against STEC during the primary production and processing of raw beef, raw milk and raw milk cheeses.

The quality of evidence varied greatly depending on study design, method of analyses, STEC serogroup used, and the scale of each study (e.g. laboratory, farm or processing plant). Many results were from laboratory or small-scale studies, which may not be scalable to meet commercial demands under a myriad of diverse production and processing conditions. There are, therefore, uncertainties as to whether these studies are truly representative of production and processing conditions, and whether the observed STEC reductions will occur in actual situations.

Scientific evaluations of intervention treatments for STEC are ideally as representative as possible of the scenario in which they would be applied; however, these studies are frequently prohibited due to the health risk associated with the introduction of a pathogen into the food manufacturing facility. Consequently, surrogate bacteria (e.g. NTS *E. coli*) are used as substitute and the results extrapolated, which means that the evidence on the effects of interventions specifically on STEC may not be available currently, or in the future. Furthermore, molecular techniques are increasingly refining current evidence, hence, existing data may be subject to future revision.

Implementing STEC monitoring plans at the farm level to measure their impact on STEC presence in raw beef and dairy products may or may not be practical because of the nature of beef and dairy production being composed of many small-scale, independent producers, with little or no integration between production phases, as well as variability in cattle STEC excretion dynamics.

Intervention strategies have generally been examined individually at specific points in the food chain. The use of multiple control measures has also been implemented sequentially on farms and in beef, ground beef and dairy processing plants. Although it is uncertain to be cumulative, the additive effect of multiple interventions applied sequentially to reduce STEC transmission in the meat or dairy production chain remains unknown, and it is almost certain that they will not completely eliminate STEC.

- Perhaps more important than the effectiveness of the intervention measures, the producers and processors making the decision to select a specific control measure must also consider the ability and logistics to install or implement the measure, its practicality, occupational health and safety concerns, environment resource management and cost.
- Beef and dairy producers and processors typically follow GAP and/or GHP to reduce the spread of pathogenic and spoilage organisms. While these practices likely reduce STEC as well, specific evidence of STEC reduction is lacking or limited.
- In the processing plant, data on the impact of interventions on quantitative reduction of STEC is limited or lacking as product inoculation studies with STEC cannot be performed in a commercial facility. As such, in-plant evidence for impact of interventions on STEC is generally based on data from prevalence studies. However, evidence on the impact of interventions on STEC obtained from research laboratories or pilot plants, can be combined with in-plant data on surrogate generic NTS *E. coli* or other microorganisms to make efficacy assessments.
- Farm based practices and interventions can reduce STEC carriage, excretion and transmission/recirculation within a herd. But these reductions can be negated at later stages of the processing chain as a result of mixing with other animals during transport and lairage. Mixing animals that have not been similarly treated will result in cross-contamination during processing.
- Good animal management and production practices include hygienic housing and bedding, low animal density, clean drinking water, biosecurity, safe and effective sanitation, and manure management. All of these will contribute to reducing faecal-oral transmission of pathogens, including STEC, among cattle.
- The impact of several dietary management and nutritional strategies on reducing STEC populations in meat and dairy animals have been explored, but with varying degree of effectiveness. There is little evidence supporting these interventions for the control of STEC.
- The use of numerous feed additives to manage STEC levels were examined. The reported effects of using probiotics, colicins, bacteriophage (in feed), and sodium chlorate *in vivo* were highly variable, depending on the agent and the

CONTROL MEASURES FOR SHIGA TOXIN-PRODUCING *ESCHERICHIA COLI* (STEC)
ASSOCIATED WITH MEAT AND DAIRY PRODUCTS

animal host. At present and based on the available evidence, these are not recommended for consideration for the control of STEC.

- Some vaccines have been shown to reduce faecal excretion of STEC O157:H7, but the efficacy is dependent on the type of vaccine and also the number of doses required.

- In the processing plant, data on the impact of interventions on quantitative reduction of STEC is limited or lacking as product inoculation studies with STEC cannot be performed in a commercial facility. As such, in-plant evidence for impact of interventions on STEC is generally based on data from prevalence studies. However, evidence on the impact of interventions on STEC obtained from research laboratories or pilot plants, can be combined with in-plant data on surrogate NTS *E. coli* or other microorganisms to make efficacy assessments.

- Long distance cattle transport increased faecal excretion and cross-contamination between animals. The exact role of lairage in spreading STEC among animals is unclear and is likely dependent on facility design, duration, stress, animal density, and cleanliness. In lairage, clean animal scoring can be used to classify clean and dirty animals, but the association between clean animal scores and reduced STEC prevalence on carcass is unclear.

- GHP measures used during processing include lairage hygiene, optimized dressing and evisceration procedures to minimize carcass contamination from the hide and gut, trimming to remove visible contamination, minimizing handling cross-contamination, and effective cooling systems to prevent microbial growth. All of these interventions contribute to reducing contamination of pathogens, including STEC, in raw beef.

- Treatments to decontaminate hides include washes, dehairing, and bacteriophage, applied before or after stunning. The reported effects were highly variable and there were practical and logistical issues for in-plant application. At present, there is limited evidence of their effectiveness in reducing transfer of STEC to carcasses

- Processing measures that specifically reduced STEC prevalence on carcasses included: steam vacuuming of visible faecal contamination on carcasses, carcass wash using hot potable water, steam pasteurization, and 24-h air chilling and combinations of these. In-plant studies showed these measures to have significant reductions in STEC prevalence.

- Despite the widespread commercial use of organic acids and other chemical agents to decontaminate pre-chill carcasses, there is wide variation in the reported reductions of STEC levels and prevalence in both research and commercial applications, depending on the trial parameters used.

- The comparative efficacy of available and putative control measures (e.g. bacteriophage, lactic acid treatments, irradiation) for reducing or eliminating STEC on primal cuts, trim and cheek meats is wide ranging and most of the studies are laboratory-based with none of those examined being performed in commercial production conditions.
- The comparative efficacy of available and putative control measures for reducing or eliminating STEC in ground beef and in retail packs was wide ranging and only high-pressure processing, gamma irradiation, and eBeam were identified as most efficacious.
- The process of grinding beef and comingling raw milk results in a broader distribution of STEC throughout the product. STEC levels in raw beef, raw milk and raw milk cheeses can vary considerably depending on primary-production, processing, and post-processing conditions, and the interventions applied.
- For raw milk, interventions using bactofugation, microfiltration, bacteriophages, eBeam and high pressure reduced bacteria, *E. coli* and/or STEC levels. But all of these interventions presented logistical issues such as the need for sub-pasteurization temperature heating, costly equipment, and may be associated with potential organoleptic changes to the product.
- For the manufacturing of raw milk cheeses, the cooking, acidification, and ripening steps, or a combination of these may be associated with STEC or *E. coli* reductions; however, the magnitude of reduction varied by STEC serotype and the type of cheese. Thus, the quality of raw milk used in cheese making along with manufacturing hurdles are crucial to reducing the risks associated with the end products.
- Some studies showed that combinations of interventions are more efficacious than individual treatments in reducing STEC levels, but the results can be inconsistent and varied depending on study parameters. Even if proven to be effective, the added cost and time for the application of combined treatments may render them impractical for use in plants.

Annexes

Primary production control strategies for STEC in beef and dairy

TYPES EVIDENCE TO SUPPORT INTERVENTION	DETAILS OF RESEARCH STUDIES	OTHER CONSIDERATIONS	DEGREE OF SUPPORT			CITATIONS
			LOW	MED	HIGH	
2.1 Animal Factors						
STEC genomics	Distinct genetic lineages of STEC are associated with human illness and colonizing cattle. Determines likelihood of strain causing human illnesses. Super shedder strains/phage types may have differences.	No specific genomic linkages are known that provide targets for intervention strategies to reduce STEC carriage by cattle or transmission. No known STEC targeting capacity.	✓			Feng *et al.*, 1998; Whittam *et al.*, 1988
2.1.1 Cattle genetics						
Host genetics	Impact both the phylogenetic diversity and the relative abundance of members of the intestinal microbiome.	Innate and acquired immunity as well as other host-microbiome communication channels may influence the establishment of STEC within the host and STEC excretion.	✓			Wang *et al.*, 2016; Munns *et al.*, 2014, 2015
Animal bread: Occurs in both dairy and beef cattle	Some evidence that hosts immune status at the recto-anal junction may influence shedding, but mechanisms are unclear. Investigation of host genomic- STEC. interactions are worthy of continued investigation.	Other factors such as animal management and diet are likely to be more important in determining shedding status than host genetics.	✓			Wang *et al.*, 2016; Munns *et al.*, 2014, 2015

TYPES EVIDENCE TO SUPPORT INTERVENTION	DETAILS OF RESEARCH STUDIES	OTHER CONSIDERATIONS	DEGREE OF SUPPORT			CITATIONS
			LOW	MED	HIGH	
Super shedder strains may have a host genetic component	Cattle that shed > 10⁴ CFU/g of faeces, play a significant role in the transmission of STEC O157:H7 within the production environment. Significant source of STEC O157:H7 during primary production.	Shedding is intermittent and not clear with regard to what controls it; could be related to sloughing of biofilms from intestinal epithelium.	✓			Wang et al., 2016; Munns et al., 2014, 2015

2.1.2 Cattle intestinal microbiome

TYPES EVIDENCE TO SUPPORT INTERVENTION	DETAILS OF RESEARCH STUDIES	OTHER CONSIDERATIONS	DEGREE OF SUPPORT			CITATIONS
			LOW	MED	HIGH	
Non-O157 STEC serogroups can be carried in ruminants as commensal-type organisms	Limited studies have indicated that STEC O157:H7 is typically the most prevalent STEC in human cases because it was the earliest to recognize and to test for. However, we now know that the prevalence of non-O157 STEC is broadly similar in cattle. Colonization with multiple serogroups simultaneously occurs. Non-O157 STEC are theorized to behave similarly to STEC O157:H7. While it appears that there are individual differences in serotype capacity to colonize the gut of cattle, there are no differences between serotypes that can be exploited by specific intervention strategies.	Regions have variable levels of non-O157/O157:H7 ratios. Insufficient information exists on carriage of O157:H7 vs non-O157 world-wide. Influence of geography, management practices, as well as animal genetic background affects prevalence. There are significant physiological differences within STEC O157/non-O157 strains that play a role in the ecological niche, and the resultant prevalence of each serotype.	✓			Arthur et al., 2002; Bonardi et al., 2004; Cernicchiaro et al., 2014; Bergholz and Whittam, 2007; Cull et al., 2017; Dewsbury, 2015; Fan et al., 2019; Free et al., 2012; Mellor et al., 2016
Evidence that phylogenetic composition of intestinal microbiota differs between super-shedders and non-shedders	No specific profiles have been attributed to super-shedders, nor have specific microbiota been shown to be associated with non-shedders. Interactions among members of the microbiome within the gastrointestinal tract merit further investigation.	Cattle gut microbiome can be altered by use of other interventions.	✓			Zaheer et al., 2017; Xu et al., 2014

TYPES EVIDENCE TO SUPPORT INTERVENTION	DETAILS OF RESEARCH STUDIES	OTHER CONSIDERATIONS	DEGREE OF SUPPORT			CITATIONS
			LOW	MED	HIGH	
2.1.3 Cattle demography						
Calf rearing practices (including bob veal)	Randomized controlled trial investigated 3 intervention packages to reduce STEC O157:H7 in young-cattle. Feeding colostrum from the calf's dam, a decrease in serum IgG conc. and high temperature-humidity index increased the likelihood of STEC O157:H7 in pre-weaned calves. Calves for veal have increased risk of shedding and carriage of STEC compared to older calves. Veal calves positive on hide and carcasses for STEC O157:H7 and non-O157 STEC. Keeping young cattle in the same groups was one the most important measures. Strategies that mitigate the effect of temperature could be advantageous.	Immunity and overall health of the calf GI tract are important factors. Questions regarding colostrum administration route (amount of colostrum ingested, passive transfer of immunity, transmission of STEC O157:H7 via colostrum, etc.). Calves housed independently or in small groups to prevent disease transmission is a GAP. Future research needed.		✓		Ellis-Iversen *et al.*, 2008; Stenkamp-Strahm *et al.*, 2018; Bosilevac *et al.*, 2017
Stage of production	Grouping cattle based on age and production status to prevent disease spread and maintain appropriate animal nutritional status.	Grouping cattle based on age and production status is a GAP.		✓		Ekong, Sanderson and Cernicchiaro, 2015; Edrington *et al.*, 2004; Venegas-Vargas *et al.*, 2016
Animal age	Calves shed STEC O157:H7 more frequently than older cattle. Cull dairy cattle often come from herds based on disease status or age, which is linked with increased carriage of STEC O157:H7.	Grouping cattle together in age-specific groups is a GAP and allows higher shedding calves to be separated from older cattle.		✓		Herriott *et al.*, 1998; Ellis-Iversen, 2008
2.2 Environmental Factors *2.2.1 Biosecurity*						
Fly control	Association of STEC O157:H7 with filth on flies and experimental transmission of STEC O157:H7 by flies.	While these effects are probably minimal in their direct impact on food safety within a farm, they represent vectors that can transfer pathogens between "clean" groups of cattle or farms. No intervention study to date.	✓			Hancock *et al.*, 1998; Talley *et al.*, 2009; Ahmad, Nagaraja and Zurek, 2007

TYPES EVIDENCE TO SUPPORT INTERVENTION	DETAILS OF RESEARCH STUDIES	OTHER CONSIDERATIONS	DEGREE OF SUPPORT			CITATIONS
			LOW	MED	HIGH	
Rodent control	STEC O157:H7 in feedlot cattle and in Norwegian rats from a large-scale farm.		✓			Čížek et al., 1999
Bird control	Wild birds as source of clonal dissemination of STEC O157:H7 among dairy farms, both migratory and native. Wild bird density & farm management are important issues.		✓			Wetzel and Lejeune, 2006; Cernicchiaro et al., 2012; Callaway, Edrington and Nisbet, 2014
Open vs closed herd (cattle population)	Randomized controlled trial that investigated three intervention packages on reduction of STEC O157:H7 in young-stock cattle farms in England & Wales. Maintaining the animals in the same groups is one of the most important measures (48 percent reduction in STEC O157:H7).	Introduction of STEC O157:H7 through incoming animals should be prevented by reducing entry to the farm. Farms with animals at pasture with water supply from natural source and with higher numbers of finishing cattle had lower prevalence.		✓	✓	Ellis-Iversen et al., 2008; Gunn et al., 2007; Garber et al., 1999; Smith et al., 2001; Sanderson et al., 2006
2.2.2. Animal density						
Animal density	Animal density linked with an increased risk of carriage of STEC O157:H7, and stocking density increased both shedding and horizontal spread of STEC O157:H7.	Density is especially important when super shedding animals are present, as density increases contact between animals. Increased density reduces environmental footprint.		✓		Frank et al., 2008; Strachan et al., 2006; Vidovic and Korber, 2006; Haus-Cheymol et al., 2006
2.2.3. Environmental hygiene						
Pen scaping	Effective in reducing presence of STEC.	Pen scraping avoids use of water flush which increases STEC prevalence.			✓	Garber et al., 1999; Smith, et al., 2002
Pen floor	Avoid muddy pen floors which favour STEC survival and spread. Effective in reducing presence of STEC.	Pen floors are more significant source of STEC O157:H7 infection than feed and water.		✓		Smith et al., 2001; Bach et al., 2005a
Animal handling facilities	Animal handling facilities including squeeze chutes (crushes) and other contact points can horizontally spread STEC on hides.	GAP that can limit cross-contamination Amount of faeces present on hide is directly related to transmission risk.		✓		Mather et al., 2007

CONTROL MEASURES FOR SHIGA TOXIN-PRODUCING *ESCHERICHIA COLI* (STEC) ASSOCIATED WITH MEAT AND DAIRY PRODUCTS

TYPES EVIDENCE TO SUPPORT INTERVENTION	DETAILS OF RESEARCH STUDIES	OTHER CONSIDERATIONS	DEGREE OF SUPPORT			CITATIONS
			LOW	MED	HIGH	
2.2.4 Manure management issues						
Solid waste composting	Composting at 55 °C to 65 °C; Covering with finished compost.Highly effective, recommended as a post treatment for manure prior to land application.	Proper carbon: nitrogen ratio required to achieve kill temperatures; temperature used as validation of the process. GAP.	✓			LeJeune et al., 2004
Slurries	Anaerobic digestion at 30–35°C with 20 days retention. Addition of lime (pH 12 for at least 2 h). Highly effective, recommended as a post treatment for manure prior to land application.	Care must be taken to monitor nutrient level and ensure that other contaminants are not released into the environment from bio-digestion of sludge. GAP.	✓			Blaustein et al., 2015
Manure applied to fields	Can serve as an STEC contaminant of both ground and surface water. Secondary treatment is recommended to reduce risk of application to both crop and forage land. Not a specific STEC intervention.	Care must be taken to monitor nutrient level and ensure that other contaminants are not released into the environment from bio-digestion of sludge. GAP.	✓			Ongeng et al., 2015
Grazing practices	Limit grazing in pastures shared with other ruminants, which can transfer STEC.	Ensure that shared pasture usage between ruminant species is limited is a GAP.	✓			Stacey et al., 2007; Duffy, 2003; Callaway et al., 2013
2.2.5 Seasonal variability and temperture						
Summer peak in excretion and prevalence	Correlation of STEC O157:H7 shedding, human cases and seasonality, weather and water. No intervention study to date.	Clear seasonality (summer peak), multiple causes: e.g. growth of STEC, growth of other vectors (protozoa), day length affecting faecal shedding, water troughs, and animal aggregation in shade and near water sources.	✓			Money et al., 2010; Ekong, Sanderson and Cernicchiaro, 2015; Besser et al., 2014; Gautam et al., 2011; Dawson et al., 2018
Temperature	STEC O157:H7 outbreak in humans after heavy rainfall (vector: sheep faeces). Manure run-off contaminating different waters such as retention ponds in feedlots, exposed surface waters, streams. No intervention study to date.	See also heat/cold stress	✓			Ogden et al., 2002; Tymensen et al., 2017; Tanaro et al., 2014; Johnson et al., 2003; Cook et al., 2011

TYPES EVIDENCE TO SUPPORT INTERVENTION	DETAILS OF RESEARCH STUDIES	OTHER CONSIDERATIONS	DEGREE OF SUPPORT			CITATIONS
			LOW	MED	HIGH	
Heat / Cold stress	Heat stress has shown limited impact on STEC shedding. Sprinkler use to alleviate heat stress demonstrated no impact on STEC O157:H7 populations. Not an intervention, but important for animal welfare and productivity.	Sprinklers can increase mud in pens which increases STEC survival and the amount of coat tag. Alleviating heat/ cold stress is animal welfare issue.	✓			Brown-Brandl *et al.*, 2009; Edrington *et al.*, 2009a

2.3 Water and feed management strategies
2.3.1 Drinking water quality and hygiene

TYPES EVIDENCE TO SUPPORT INTERVENTION	DETAILS OF RESEARCH STUDIES	OTHER CONSIDERATIONS	DEGREE OF SUPPORT			CITATIONS
			LOW	MED	HIGH	
Cleaning of water troughs	Water troughs are sources of STEC O157. Higher cleaning rate predicted as being particularly efficacious at reducing the load of STEC at the farm and at increasing death rate of STEC O157:H7. Cleaning of water troughs predicted to reduce STEC population and dissemination of STEC in cattle production.	No effect found on chlorinated versus non chlorinated water on prevalence on STEC O157:H7. And no effect of improved water hygiene (randomized controlled trial). Some limitations to its effectiveness based on environmental conditions.	✓			VosoughAhmadi *et al.*, 2007; Ayscue *et al.*, 2009; Ellis-Iversen *et al.*, 2008; LeJeune *et al.*, 2004
Water-to-cattle ratio (automatic refilling water troughs)	Keeping water levels high in water troughs suggested to reduce prevalence of STEC O157:H7. Association between reduced water level and increased STEC O157:H7 prevalence.	Dilution of pathogens; dependent on water availability/cost and can increase water utilization by a farm.	✓			Beauvais *et al.*, 2018

2.3.2 Drinking water treatment

TYPES EVIDENCE TO SUPPORT INTERVENTION	DETAILS OF RESEARCH STUDIES	OTHER CONSIDERATIONS	DEGREE OF SUPPORT			CITATIONS
			LOW	MED	HIGH	
Chlorine	Chlorine is an effective disinfectant and can be used to clean water troughs, but troughs quickly become re-contaminated after cleaning.	Antimicrobial activity of chlorine is reduced if high levels of organic matter come in contact with chlorinated water or if it is exposed to UV light.	✓			LeJeune, Besser and Hancock, 2001, 2004; Smith *et al.*, 2002;
EO water	An effective disinfectant and possibly more active than just chlorinated water, but subject to same inactivation by UV light and organic matter.	Has been used in processing and postprocessing environments.	✓			Bosilevac *et al.*, 2005a; Stevenson *et al.*, 2004

TYPES EVIDENCE TO SUPPORT INTERVENTION	DETAILS OF RESEARCH STUDIES	OTHER CONSIDERATIONS	DEGREE OF SUPPORT			CITATIONS
			LOW	MED	HIGH	
2.3.3 Diet composition, feeding strategies and feed hygiene						
Cleaning of feed troughs	Feed hygiene to reduce STEC faecal contamination. No relationship between pens shedding STEC O157:H7 and recovery from feed (and water), but modelling suggests an influential source.	Providing clean water and feed is a GAP and is critical to ensuring good animal health and productivity.	✓			Smith *et al.*, 2001; Dodd *et al.*, 2003; Sanderson *et al.*, 2006; Ayscue *et al.*, 2009; Berry and Wells, 2010
Forage: concentrate ratio: Variation in diet composition in regard to forage to concentrate ratio	*E. coli* populations (both generic and STEC) are generally higher in grain-fed cattle than in forage fed. Experimentally inoculated calves were fed high grain or high forage diets on the duration or shedding of faecal STEC O157:H7 populations in experimentally inoculated calves have found that low quality forage feeding caused a faster rate of death of STEC O157 populations in manure. Faeces from grain fed cattle had higher VFA concentrations and lower pH allowing STEC O157:H7 populations to survive longer than in faeces from grass-fed cattle. Controlled studies have been few and mostly observational.	Feeding forage diets to all cattle would reduce the availability of animal protein and is difficult to implement in feedlots and arid regions. Effect also appears to be linked to forage quality (tentative). However, the host/dietary/microbial factors underlying the "super shedder" status of cattle remains unknown, as do factors that allow simple gut colonization by STEC O157:H7. A better understanding of microbial populations and physiology of the gastrointestinal tract of cattle will allow reduction of STEC O157:H7 at pre-harvest through diet.	✓	✓		Callaway *et al.*, 2009, 2013
Rapid Dietary shift from grain to forage	A rapid shift from a high grain to hay diet resulted in a 3 $\log_{10}$ reduction in NTS *E. coli* populations.	Dietary shifts are difficult to implement in feedlots and in arid regions, with logistical challenges abound.	✓			Callaway *et al.*, 2009, 2013; Diez Gonzalez *et al.*, 1998
Grain type	Barley feeding is linked with increased shedding and survival of STEC O157:H7 in faeces compared to corn. Barley doubled the survival time of STEC O157:H7 in faeces and shedding concentration of STEC O157:H7 < 0.5 $\log_{10}$ vs corn. Barley feeding also resulted in an increase in excreted STEC populations.	Barley is a grain that is often fed to cattle which is more rapidly fermented in the rumen than corn, and so, little barley passes to the hindgut and little starch reaches the lower gut.			✓	Bach *et al.*, 2005a, 2005b, 2002; Berg *et al.*, 2004

TYPES EVIDENCE TO SUPPORT INTERVENTION	DETAILS OF RESEARCH STUDIES	OTHER CONSIDERATIONS	DEGREE OF SUPPORT			CITATIONS
			LOW	MED	HIGH	
Processing of grain (cracking, steam flaking, etc.)	Dry rolling of grain increases starch flow to the hind gut and reduces STEC O157:H7 populations compared to Steam flaked grain. Steam flaking corn increased STEC O157:H7 shedding in heifers compared to whole corn. Faecal starch concentration and pH are not linked to STEC O157:H7 shedding.	Location of fermentation of starch in cattle is shifted by grain processing, which impacts animal health and growth efficiency. Faecal starch concentration and pH were not linked to STEC O157:H7 shedding, yet post-ruminal starch infusion increased NTS *E. coli* populations in the hindgut.		✓		Fox *et al.*, 2007; Depenbusch *et al.*, 2008
Dried Distiller's Grains with solubles (DDGS); Wet Distiller's Grains with Solubles (WDGS)	Feeding 40% WDGS increased faecal shedding of STEC O157:H7 (10% in 0 vs 70%), but 15% DGS did not increase excretion. Experimentally inoculated cattle showed that STEC O157:H7 manure populations were decreased from 6.28–2.48 $\log_{10}$ CFU/g by 40% WDGS. DGS fed at 20, 40, and 60% corn WDGS increased STEC survival in manure from 1-3 $\log_{10}$ CFU/g of faeces. Feeding 40% DGS increased faecal STEC populations by > 3 $\log_{10}$, and increased survival time in manure. Evidence supports recommending feeding Distillers or Brewer's grains at levels < 15% is thought to not increase STEC O157:H7 populations.	Impact of including drieddistillers grains is highly variable due to poor QC/QA. Effects of feeding >40% DGS increases risk of faecal STEC shedding.		✓		Jacob *et al.*, 2008, 2009; Wells *et al.*, 2011; Paddock *et al.*, 2013; Berry *et al.*, 2017
Tannins	Feeding of Tasco-14 (Phlorotannin-containing) displayed anti-STEC activity higher than other terrestrial tannin sources.	Complete or partial elimination of different STEC-Serotypes, with STEC O157:H7 reduced by up to 36%.	✓	✓		Zhou *et al.*, 2018; Braden *et al.*, 2004
Essential oils	A large variety of diverse feed additives can be fed to ruminants (e.g. feeding orange peel to sheep; thyme). Breadth of category limits the ability to determine effectiveness of individual essential oils.	Conflicting or not repeatable study results done in small scale challenge studies.	✓			Callaway *et al.*, 2011; Jacob, Callaway and Nagaraja, 2009

CONTROL MEASURES FOR SHIGA TOXIN-PRODUCING *ESCHERICHIA COLI* (STEC) ASSOCIATED WITH MEAT AND DAIRY PRODUCTS

TYPES EVIDENCE TO SUPPORT INTERVENTION	DETAILS OF RESEARCH STUDIES	OTHER CONSIDERATIONS	DEGREE OF SUPPORT			CITATIONS
			LOW	MED	HIGH	
2.3.4 Feed additives						
Priobiotics/Direct fed microbials (DFM)	Some DFM are effective in reducing prevalence of STEC O157:H7 faecal shedding in beef cattle. Most efficient: *Lactobacillus acidophilus* (NP51) and *Propionibacterium freudenreichii* (NP24); at doses of 10^9 CFU/animal/day, for 137 days reduced prevalence of STEC O157 faecal sheding in beef cattle. STEC O157:H7 isolation was 74% less likely on hides and in faeces.	DFM (eubiotic and post-biotic) effectiveness vary widely based on the active organism in the product as well as dosing level.		✓		Wisener *et al.*, 2015; Stephens *et al.*, 2007
Competitive exclusion cultures (calves)	Oral administration of probiotic *E. coli* to calves. Effectiveness shown against STEC O26 & O111, but not O157.		✓			Zhao *et al.*, 2003
Phage hijack cellular metabolic machinery of bacteria and cause bacterial lysis	Efficacy in primary production has been mixed, with either little change in STEC numbers or the establishment of a cyclic response where STEC counts decline with phage numbers increase, but phage numbers decrease and STEC counts increase. Phage show promise for use in processing and post-processing phases.	Phage are often highly specific with most having a narrow host range, which is limited to one or a few STEC serotypes. STEC can also become resistant to individual phage.	✓			Sabouri *et al.*, 2017; Liu *et al.*, 2015; Arthur *et al.*, 2017; Wang *et al.*, 2017
Colicins kill STEC O157:H7	Colicins reduced STEC O157:H7 strains in vitro and in small ruminants but do not differentiate between generic and STEC. Colicin production by plants allows for production scale up for inclusion in rations. Colicin producing probiotic cultures have been used to reduce STEC O157:H7 in cattle in experimental studies.	Evidence of colicin anti-STEC activity has been more promising in ground beef and in cleaning processing facility surfaces.	✓			Callaway *et al.*, 2004; Schulz *et al.*, 2015; Schamberger *et al.*, 2004

TYPES EVIDENCE TO SUPPORT INTERVENTION	DETAILS OF RESEARCH STUDIES	OTHER CONSIDERATIONS	DEGREE OF SUPPORT			CITATIONS
			LOW	MED	HIGH	
Sodium chlorate	*E. coli*, including STEC can respire anaerobically using the enzyme nitrate reductase, which also reduces sodium chlorate to chlorite, an antimicrobial compound. Feeding Sodium chlorate to sheep reduced inoculated STEC O157:H7 in the rumen, cecum, and colon by 1-4 $\log_{10}$. Treatment in cattle reduced inoculated STEC O157:H7 populations by 2-3 $\log_{10}$ throughout the gut with no impact on meat quality.	Product has been developed but not on the market pending on the approval process, which can be lengthy.		✓		Edrington *et al.*, 2004; Callaway *et al.*, 2002, 2003

2.4 Vaccines and clinical antimicrobials
2.4.1 Vaccines

Novel bivalent vaccine against STEC infection via *Clostridium perfringens* enterotoxin (CPE)-based protein engineered for vaccine design and delivery system	Administration of C-terminus of CPE (C-CPE) alone to mice induces C-CPE-specific IgM, but not IgG response due to its low antigenicity. In contrast, administering Stx2B–C-CPE, sufficient IgG immune responses with neutralizing activity against CPE were induced. Formulations against STEC strains are both protective in mice. However, mice are not an effective model for ruminants.	The C-CPE is non-toxic and is the part of the toxin that binds to epithelial cells via the claudins in tight junctions; however, C-CPE has low antigenicity.	Low to high, depending upon vaccine			Lan, Hosomi and Kunisawa, 2019
Genetically inactivated recombinant Shiga toxoids (rStx1MUT/ rStx2MUT)	A group of 24 calves was passively (fed colostrum from immunized cows) and actively (intra-muscularly at 5th and 8th week) vaccinated. Another 24 calves served as unvaccinated controls (fed with low anti-Stx colostrum or injected with placebo). Each group was divided according to the vitamin E concentration (moderate and high supplemented) they received by milk replacer.	The effective transfer of Stx-neutralizing antibodies from dams to calves via colostrum was confirmed by Vero cell assays. Serum antibody titers in calves differed significantly between the vaccinated and the control group until the 16th week of life.	Low to high, depending upon vaccine			Schmidt *et al.*, 2018

TYPES EVIDENCE TO SUPPORT INTERVENTION	DETAILS OF RESEARCH STUDIES	OTHER CONSIDERATIONS	DEGREE OF SUPPORT			CITATIONS
			LOW	MED	HIGH	
Evaluation of biological safety *in vitro* and immunogenicity *in vivo* of recombinant *Escherichia coli* Shiga toxoids as candidate vaccines in cattle	The experiment used two conventionally raised bull calves aged 11 months that tested negative for Stx-specific antibodies (16 and 4 weeks before the trial by VNA).	In cattle, Stx suppresses the immune system thereby promoting long-term STEC shedding. First infections of animals at calves' age coincide with the lack of Stx-specific antibodies. Antibodies in sera of cattle naturally infected with STEC recognized the rStx mut toxoids equally well as the recombinant wild type toxins.	Low to high, depending upon vaccine			Kerner *et al.*, 2015; Schmidt *et al.*, 2018
Immune response in calves vaccinated with type three secretion system Antigens and Stx 2B Subunit of STEC O157:H7	Calves were tested for STEC shedding 16 weeks before the trial then vaccinated with two doses of different vaccine formulations: two antigens (IntiminC280, EspB), three antigens (IntiminC280, EspB, BLS-Stx2B), BLS-Stx2B alone and a non-vaccinated group as control.	All antigens were expressed as recombinant proteins in *E. coli*. Specific IgG titer increased in vaccinated calves and the inclusion of BLS-Stx2B in the formulation seemed to have a stimulated the humoral response to IntiminC280 and EspB after the booster.	Low to high, depending upon vaccine			Martorelli *et al.*, 2017
Outer membrane vesicle (OMV)-based vaccine formulations against STEC are both protective in mice and immunogenic in calves	Fifteen calves from a beef producing brand and between six and eight months old were allocated to a single pen and randomly divided into three groups of five. Group 1 was assigned a 50 µg OMVi plus aluminum adjuvant per dose, group 2 a 100 µg OMVi plus aluminum adjuvant per dose, and group 3 was treated with aluminum adjuvant in saline (control). Each group was vaccinated subcutaneously on days 0, 21, and 42.	OMV obtained after detergent treatment of gram negative bacteria have been used for decades for producing many licensed vaccines. These nanoparticles are not only multi-antigenic in nature but also potent immunopotentiators and immunomodulators. Formulations based on chemical inactivated OMV (OMVi) obtained from a STEC O157:H7 strain (was found to protect against pathogenicity in a murine model and to be immunogenic in calves. These initial studies suggest that STEC-derived OMV has potential to be developed as both human and veterinary vaccines.	Low to high, depending upon vaccine			Fingermann *et al.*, 2018

TYPES EVIDENCE TO SUPPORT INTERVENTION	DETAILS OF RESEARCH STUDIES	OTHER CONSIDERATIONS	DEGREE OF SUPPORT			CITATIONS
			LOW	MED	HIGH	
Commercial vaccine products	Econiche™, designed to reduce the shedding of STEC O157:H7 by cattle, has received full licensing approval from the Canadian Food Inspection Agency (CFIA). Econiche is now available for unrestricted use by Canadian cattle growers but there are resistance concerns and no proven efficacy has been demonstrated in scientific studies. Due to valid concerns regarding the use of antimicrobials in animal husbandry, some countries have banned the use of antimicrobials growth promoters in cattle farming.	Although this vaccine induces antibody responses effective in significantly reducing colonization, they are only partially protective.	Low to high depending upon vaccine			Snedeker, Campbell and Sargeant, 2012

2.4.2 Clinical antimicrobials

TYPES EVIDENCE TO SUPPORT INTERVENTION	DETAILS OF RESEARCH STUDIES	OTHER CONSIDERATIONS	DEGREE OF SUPPORT			CITATIONS
			LOW	MED	HIGH	
• Ionophores, tylosin, chlor-tetracycline, and oxytetra-cycline-feedlot cattle (subtherapeutic levels) • Tylosin-prevent hepatic abscessation and promote growth • Chlortetracycline and oxytetracycline: used at therapeutic levels • Other	The efficacy of these antimicrobials against STEC has not been shown *in vivo* or *in vitro* studies. But one would expect that the use of broad spectrum or bacteriostatic and/or bactericidal antimicrobials against gram negative bacteria, would most probably reduce STEC colonization and shedding. But the use of antimicrobials remains controversial because of antimicrobial resistance concerns and they can induce Stx bacteriophage which may spread Stx-encoding genes to naive *E. coli*. No proven efficacy of use has been demonstrated in scientific studies. Due to valid concerns on their use, some countries have banned the use of antimicrobials growth promoters in cattle farming.	Key considerations: These antimicrobials 1) may induce Stx bacteriophages that are able to transduce stx-encoding genes and antimicrobial resistance genes to naive *E. coli*, thereby expanding the STEC pool in individual animals and at the herd level. (*in vitro* studies). 2) Some antimicrobials may exert selective pressure on intestinal microbiota, thereby favouring the survival of antimicrobial resistant STEC (no studies). 3) Antimicrobials against gram positive bacteria may favour the dominance of gram-negative bacteria including STEC or vice versa.	✓	✓		Allison, 2007; Köhler, Karch and Schmidt, 2000; Kimmit *et al.*, 2000; Herold *et al.*, 2004; Colavecchio *et al.*, 2017; USDA/FSIS, 2014
Neomycin sulfate	Reduced STEC shedding in cattle.	Antimicrobial Resistance issues are significant and preclude recommendation.	✓			Elder *et al.*, 2002

TYPES EVIDENCE TO SUPPORT INTERVENTION	DETAILS OF RESEARCH STUDIES	OTHER CONSIDERATIONS	DEGREE OF SUPPORT			CITATIONS
			LOW	MED	HIGH	
Ionophores	Ionophores are fed to improve ruminant feed efficiency. It alters the microbial population of the rumen through inhibition of gram-positive bacteria, resulting in a change in fermentation end products and reduction in methane production. Highly successful in improving feed efficiency and reducing environmental pollution. Theoretical competitive advantage to Gram negative species like STEC, butnotshown to occur in sheep or cattle. Percent of animals shedding STEC O157:H7 was greater for monensin fed steers than for controls, yet the presence of monensin, did not affect the percentage of animals in the pen shedding STEC O157:H7.	Antimicrobial usage concerns. Not an intervention directly against STEC. Inclusion of ionophores in cattle feed is included to improve animal growth efficiency.	✓			Edrington et al., 2003; McAllister et al., 2006; Hales et al., 2017

2.4.3 Beta-agonists/hormones

Ractopamine and zilpaterol	Theorized to increase STEC shedding as a result of increased stress on the animal and possibly promoting the growth of STEC, however no impact has been demonstrated.	Zilpaterol is no longer used, as no evidence from research studies showed that beta-agonists alter shedding of STEC O157:H7.	✓			Wells et al., 2017; Paddock et al., 2011; Edrington et al., 2009b

2.5 Dairy production specific interventions

Milking environment	Certain hygiene interventions on the milking equipment and the environment in a Flemish dairy herd were associated with a decrease in bacterial counts.	The consistent application of a few hygiene practices could significantly improve the microbiological quality of milk.	✓			Verbeke et al., 2014

TYPES EVIDENCE TO SUPPORT INTERVENTION	DETAILS OF RESEARCH STUDIES	OTHER CONSIDERATIONS	DEGREE OF SUPPORT			CITATIONS
			LOW	MED	HIGH	
Udder hygiene	Washing teats with a sanitizer compared with no treatment reduced microbial load by 44%, whilst washing with a sanitizer and drying decreased microbial load by 85%. Washing of teats with an effective disinfectant (chlorine) and then drying was the most effective in another study. Using an automated teat scrubber with chlorine dioxide disinfection and drying also effectively reduced bacterial loads on teats. Overall, pre-dipping teats into disinfectant followed by drying has been shown as effective means of teat skin sanitation.	Lack of correlation between cleaning regime and total viable counts, *Enterobacteriaceae* or *E. coli* levels in milk were reported. The environment, milking equipment and water are also contributors of milk contamination.		✓		Galton, Petrsson and Merrill, 1986; Fremaux *et al.*, 2006; Gibson *et al.*, 2008; Elmoslemany *et al.*, 2010; Baumberger, Guarín and Ruegg, 2016
Hygienic storage of milk	Reducing contamination and opportunity for growth of spoilage and pathogenic organisms is an important GHP.	Proper storage temperature and hygiene reduces or prevents growth of organisms, including STEC.	✓			
2.6 Animal transportation						
Feed withdrawal prior to slaughter	Feed withdrawl for >8 h prior to animal slaughter reduces gut fill (and risk of gut rupture during subsequent evisceration) and reduces production of faeces that can be spread to the hide or carcass during processing. Reducing feed in the gut results in a reduction of volatile fatty acids (VFA) production in the gut. VFA concentrations inhibit the growth of STEC and *Salmonella*.	Fasting increases prevalence and concentration of STEC in cattle; same observed in feeding low quality forages. Fasting cattle is a GHP that reduces gut rupture and faecal contamination of hides.	✓			Jordan *et al.*, 1998; Buchko *et al.*, 2000; Hovde *et al.*, 1999; Pointon, Kiermeier and Fegan, 2012

TYPES EVIDENCE TO SUPPORT INTERVENTION	DETAILS OF RESEARCH STUDIES	OTHER CONSIDERATIONS	DEGREE OF SUPPORT			CITATIONS
			LOW	MED	HIGH	
Transport distance and/or duration	Association between STEC shedding with distance of transport have been inconsistent.		✓			Bach et al., 2004; Arthur et al., 2007a; Dewell et al., 2008; Stanford et al., 2011; Brown-Brandl et al., 2009; Schuehle-Pfeiffer et al., 2009
Trailer hygiene	Potential source of contamination, similar to animal handling facilities.	No association measured.	✓			Arthur et al., 2007a
Plane of nutrition prior to slaughter	Cattle undergoing negative energy balance are subject to metabolic disorders. The role of plane of nutrition on STEC colonization is not clear.	No association measured.	✓			Callaway et al., 2009

Annex 2

Processing control strategies for STEC in beef

TYPES EVIDENCE TO SUPPORT INTERVENTION	DETAILS OF RESEARCH STUDIES	OTHER CONSIDERATIONS	DEGREE OF SUPPORT			CITATIONS
			LOW	MED	HIGH	
3.1 Lairage						
Logistic scheduling for slaughter	Younger animals have a higher risk for STEC carriage. Calves for veal have increased risk of shedding and carriage of STEC compared to older calves. Keeping young cattle in the same groups was identified as one the most important measures.	GHP; keep higher shedding calves to be separated from older cattle.	✓			Ellis-Iversen et al., 2008
Logististical slaughter based on hide cleanliness	Association between hide cleanliness scores and prevalence of pathogenic bacteria has not been consistently reported. Hide cleanliness scores have been significantly associated with STEC O157:H7 after controlling for season; specific seasons and regions have been associated with hide contamination of different STEC serotypes in a specific locality.	GHP; supplemental pre-harvest interventions could be considered in seasons or regions of known higher prevalence.	✓			FAO and WHO, 2005; Cernicchiaro et al., 2020; Schneider et al., 2018; van Donkersgoed et al., 1997; Brown et al., 2000; Keen & Elder, 2002; Smith et al., 2005b; Nastasijevic, Mitrovic and Buncic, 2008; Antic et al., 2010a, 2010b; Blagojevic et al., 2012
Lairage cleanliness	Pressure washing with water, quaternary ammonium chloride or steam resulted in 0.9–5.8 $\log_{10}$ CFU/cm² reduction in *E. coli* and *Enterobacteriaceae*.	GHP; no specific data on STEC.	✓			Small et al., 2007

TYPES EVIDENCE TO SUPPORT INTERVENTION	DETAILS OF RESEARCH STUDIES	OTHER CONSIDERATIONS	DEGREE OF SUPPORT LOW	MED	HIGH	CITATIONS
Livestock cleanliness	The cleanliness of the lairage environment is important in the maintenance of coat cleanliness. Measures include washing trailers, cattle handling facilities, holding pens between uses, regularly removed pen floor faecal material. Different cleaning and disinfection procedures might be used, and a key issue might be drying the lairage pens after cleaning and disinfection. Lots of cattle held in STEC O157:H7-positive lairage pens had 8 times greater risk of having positive slaughter hide samples (RR=8.0; 95%CI =1.6–38.8). Transport and lairage do not cause an increase in the prevalence of STEC O157:H7 faecal shedding in cattle, as demonstrated by a higher prevalence of STEC O157:H7 shedding in cattle sampled on the farm than during post-transport or lairage.	GHP; increased water usage; water used should be fit-for-purpose.	✓			Avery et al., 2002; Arthur et al., 2008; Mather et al., 2008; Dewel et al., 2008; Small et al., 2003; Minihan et al., 2003; Walia et al., 2017; FAO and WHO, 2005
Holding animals in lairage	Withdrawal feed up to 12 h while in lairage pens could reduces faecal output and soiling of environment and hide. Water misting animals in holding pens.	GHP; improves dressing percentage.	✓			

3.2 Hide decontamination
3.2.1 Bacteriophage

Bacteriophage	Lab experiment on hide treated with two phages (e11/2, e4/1c) showed a significant reduction of STEC O157:H7 after 1 h. Spray application in lairage – did not produce a significant reduction in levels or prevalence of STEC O157:H7. Lab experiment on hide treated with phage cocktail to several STEC serotypes for 1 h. A reduction was observed, but not at a high efficacy.	Efficacy studies are lacking; Highly adopted in USA in warm months;varying regulatory issues; cost relatively low.	✓			Coffey et al., 2011; Arthur et al., 2017; Tolen et al., 2018

3.2.2 Hide wash with ambient or hot water, organic acid and other chemicals

Ambient water	Washing with power hose for 3 min removed faecal contamination and STEC O157:H7 inoculated onto hide. Conflicts with the need for dry animals.	Extra water usage can be costly. Conflicts with the need for dry animals.	✓			Byrne et al., 2000
Ozonated and electrolyzed oxidizing water	Reduced EB counts by 3.4 and 4.3 log_{10} CFU/100 cm^2, respectively.	Not done in commercial conditions.	✓			Bosilevac et al., 2005a

TYPES EVIDENCE TO SUPPORT INTERVENTION	DETAILS OF RESEARCH STUDIES	OTHER CONSIDERATIONS	DEGREE OF SUPPORT			CITATIONS
			LOW	MED	HIGH	
Other chemicals	A hide wash cabinet (water and chlorine 100–200 ppm spray at the end) to be used in small and medium size plants Reduced STEC O157:H7 prevalence on hides from 35 to 13%. Cattle prewashed with water a day before harvest, then immediately before stunning, they were sprayed twice with 1% CPC for 3 min and then 1 min; CPC reduced prevalence of STEC O157:H7 by 18%. The use of an in-line hide-wash cabinet that used a sodium hydroxide wash and a chlorinated (1 ppm) water rinse. Hides sampled before entering and after exiting the cabinet had APC and EB counts that were reduced by 2.1 and 3.4 $\log_{10}$ CFU/100 cm^2, respectively, and the prevalence of STEC O157:H7 on hides was reduced from 44–17%. Whole beef hides were inoculated with STEC O157:H7 and decontaminated with spray solutions of sodium hydroxide (1.5%) followed by high-pressure washing with chlorinated (0.02%) water (SHC; both applied at 23°C), potassium cyanate (PC; 2.4%, 30oC) or sodium sulfide (SS; 6.2%, 30oC).Resulted in the greatest reductions of STEC O157:H7 (P < 0.05), by 5.1, 4.8 and 5.0 $\log_{10}$ CFU/cm^2, respectively. Hide pieces were treated with 1% caprylic acid (CA) and 1% β-resorcylic acid (BA) applied at 23°C and 60 °C sampled after 2 and 5 min. All treatments more effective at 60 °C, but in general 3-4 $\log_{10}$ reduction. No "real world" studies at slaughter houses. Study on pre-slaughter wash; I) Single water wash (1,325 l); II) Lactic acid (0.5 ± 0.2%); III) Double water wash; IV) Chlorine (50 ppm). Each wash lasted for 30 s. Increase in aerobic plate counts, coliforms, and *E. coli*.	Animal welfare issues for use on live animal hides. Potential environmental issues with disposal and practical use in plant.	✓			Brown *et al.*, 2000; Bosilevac *et al.*, 2004; Bosilevac *et al.*, 2005b; Bosilevac *et al.*, 2006; Carlson *et al.*, 2008; Mies *et al.*, 2004

TYPES EVIDENCE TO SUPPORT INTERVENTION	DETAILS OF RESEARCH STUDIES	OTHER CONSIDERATIONS	DEGREE OF SUPPORT			CITATIONS
			LOW	MED	HIGH	
Washing using water	Hide was cabinet - washing with water (cold) + chlorine spray (100-200 ppm). Wash with water followed by water rinsing with subsequent vacuuming reduced bacterial load. Ozonated and electrolyzed water - systematic literature review and meta-analysis. Effect size or intervention effectiveness was measured as raw $\log_{10}$ reduction, least-squares means were calculated.	Only slightly decreased prevalence of STEC O157:H7. But also, reduced STEC O157:H7 load, meaning the enumeration data indicated that the hide cabinet was effective. Least-squares mean reductions in $\log_{10}$ CFU/cm^2 on hide surfaces (n = 47), 0.08 [95%CI, 0.94-1.11] for water wash. Redistribute microbial contamination (forequarter sites).	✓			Arthur et al., 2007b; Bosilevac et al., 2005a; Bosilevac et al., 2005b; Zhilyaev et al., 2017
Organic acids	Least-squares mean reductions ($\log_{10}$ CFU/cm^2). On hide surfaces (n = 47), least-squares mean reductions were 2.21 [95%CI, 1.36–3.05] for acetic acid, 3.02 [95%CI, 2.16–3.88] for lactic acid. Systematic literature review and meta-analysis. Effect size or intervention effectiveness was measured as raw $\log_{10}$ reduction. Least-squares means were calculated.	May select for acid-resistant bacteria; Increase equipment corrosion; Environmental and safety of employee issues.	✓			
Other chemicals	1.6% Sodium hydroxide or 4% trisodium phosphate or 4% chlorofoam or 4% phosphoric acid. Rinse with water or acidified chlorine. Acidified chlorine (sodium hypochlorite with acetic acid) cetylpyridinium chloride (CPC). Hypobromous acid, reduced APC, TTC and EC by 2–3.8 $\log_{10}$. Least-squares mean reductions ($\log_{10}$ CFU/cm^2). On hide surfaces (n = 47), least-squares mean reductions were, 3.66 [95%CI, 2.60-4.72] for sodium hydroxide.	May select for acid-resistant bacteria; Increase equipment corrosion; Environmental and safety of employee issues.	✓			Bosilevac et al., 2004, 2005b; Schmidt et al., 2012; Zhilyaev et al., 2017
3.2.3 Hide clipping, coating and chemical dehairing						
Hide clipping, coating and chemical dehairing	Food-grade resin in ethanol (Shellac) reduced hide STEC O157:H7 by 3.7 $\log_{10}$ CFU/cm^2. Laboratory-based study of inoculated hides. In a small commercial abattoir under "worst-case" conditions (slaughtering dirty cattle, inadequate process hygiene), treatment of hides with Shellac reduced 1.7 $\log_{10}$ CFU/cm^2, 1.4 $\log_{10}$ CFU/cm^2 and 1.3 $\log_{10}$ CFU/cm^2 of TVC, EC and GEC, respectively.	Not commercially available and no data from commercial scale studies.	✓			Antic et al., 2012, 2010b

TYPES EVIDENCE TO SUPPORT INTERVENTION	DETAILS OF RESEARCH STUDIES	OTHER CONSIDERATIONS	DEGREE OF SUPPORT			CITATIONS
			LOW	MED	HIGH	
Hide de-hairing	Chemical dehairing; 10% sodium sulphide, water washes, 3% hydrogen peroxide reduced visible contamination but did not reduce total coliform counts, APC or EC. Study on small hide pieces (controlled lab conditions). Study of conventional and chemical dehairing Identified significant reduction of bacterial load (aerobic counts, coliforms, *E. coli*) and a reduction in STEC O157:H7 of 5 $\log_{10}$ CFU/cm^2 on inoculated hides.	Required cabinet; waste management (sodium sulphide); employees health and safety.	✓			Schnell *et al.*, 1995; Castillo *et al.*, 1998a; Nou *et al.*, 2003
Clipping to remove faecal material	Clipping hair from hides and singeing with handled blowtorch. Clipping followed by application of 1% CPC. Chemical dehairing resulted in lower bacterial load (~2 $\log_{10}$) and reduced prevalence of STEC O157:H7. Reduced total viable bacteria by 2.3 $\log_{10}$ CFU/cm^2. Produced the greatest reduction of APC (3.8 $\log_{10}$) on the hide surface.	Not conducted under commercial conditions.	✓			Small, Wells-Burr and Buncic, 2005; Baird *et al.*, 2006

3.3 Slaughter and dressing

TYPES EVIDENCE TO SUPPORT INTERVENTION	DETAILS OF RESEARCH STUDIES	OTHER CONSIDERATIONS	DEGREE OF SUPPORT			CITATIONS
			LOW	MED	HIGH	
Speed of processing	The speed at which animals are moved along the processing line has been reported to have both positive and negative effects on TVCs on carcasses following dressing.	Evidencs is inconsistent.	✓			Sheridan, 1998
Hide removal	Either upward or downward hide pulling system. Total viable counts (TVCs) on forequarter (3 cm^2) (n=15) indicated carcass contamination using a downward pulling system resulted in significantly lower TVC (0.4 $\log_{10}$ CFU/cm^2) than an upward pulling system (1.2 $\log_{10}$ CFU/cm^2). No significant difference in total carcass contamination (8 sites, n=36) with TVC and *Enterobacteriaceae* on specific sites (flank, shin, brisket, neck); thought to be due to GHP and not direction of hide pulling.	Evidence is inconsistent.	✓			Kang *et al.*, 2019; Kennedy, Giotis and McKevitt, 2014

TYPES EVIDENCE TO SUPPORT INTERVENTION	DETAILS OF RESEARCH STUDIES	OTHER CONSIDERATIONS	DEGREE OF SUPPORT			CITATIONS
			LOW	MED	HIGH	
Pre- and Evisceration Processes	Bagging or tying of the bung: Intestines should not be severed from the stomach during evisceration and no other opening should be made into an intestine, unless the intestines are first effectively tied to prevent spillage; except in the case of poultry and game birds. A commercial facility study showed that bunging before as compared with after the pre-evisceration wash, resulted in lower contamination rates of carcasses with STEC (35% vs. 58.3%) and STEC O157:H7 (1.5 vs. 5%).	GHP		✓		Greig *et al.*, 2012; Sheridan, 1998; Stopforth *et al.*, 2006
Removal of visible faecal material from carcass	Knife and steel sterilized by immersion in a thermostatically controlled water-bath at 82 °C for at least 30 s or an equivalent combination to result in a 2 $\log_{10}$ reduction of *E. coli*. Using clean disposable gloves and sterilized knives and steels resulted in s ignificant differences between TVCs on the brisket though not on the hocks. Lower temperature and longer time combinations could provide equivalent $\log_{10}$ reductions to meet industry requirements and using a 2-knife system could overcome delays for workers.	GHP; hot water use is a potential health and safety issues for employees.		✓		McEvoy *et al.*, 2001; Eustace *et al.*, 2007; Goulter, Dykes and Small, 2008
Trimming	Knife trimming can result in higher $\log_{10}$ reductions than a water wash and these can be reduced by combinations with other interventions. Reductions in $\log_{10}$ CFU/cm² in inoculation experiments were: TVC - 4.3, STEC O157:H7 - 3.1, *Enterobacteriaceae* - 4.1, *E. coli* - 4.1. In a commercial plant, reduction in TVCs was 3 $\log_{10}$ CFU/cm². In 24 commercial plants, trimming alone reduced TVCs by 0.44 $\log_{10}$ CFU/cm² and prevalence of *E. coli* by 29.1%; trimming plus another intervention (hot water, lactic acid, steam vacuum) reduced TVCs by 0.61 $\log_{10}$ CFU/cm² and prevalence of *E. coli* by 36.8%. Additional interventions did not reduce *Salmonella* prevalence.	Redistribution of bacterial contamination on carcass from and cross-contamination. Effectiveness can depend on employee skill.		✓		Horchner *et al.*, 2020; Castillo, 1998b; Prasai, *et al.*, 1995

TYPES EVIDENCE TO SUPPORT INTERVENTION	DETAILS OF RESEARCH STUDIES	OTHER CONSIDERATIONS	DEGREE OF SUPPORT			CITATIONS
			LOW	MED	HIGH	
Steam vacuuming	Steam vacuum applies steam or hot water (approx. 82 °C – 95 °C) using spray nozzles to loosen visible soil and inactivate bacteria and a vacuum to remove contaminants. Commercial steam vacuum systems have been reported to reduce *E. coli* between 3.0-5.5 $\log_{10}$ CFU/100 cm^2. In a meta-analysis, the mean $\log_{10}$ reduction of *E. coli* was 3.09 $\log_{10}$ CFU/cm^2 and thwas a higher reduction than from water or organic acid washes. On naturally contaminated carcasses under commercial conditions, steam vacuuming after carcass trimming was reported to reduce mean TVCs between 0.4-0.9 $\log_{10}$ CFU/cm^2 and increased distribution of TVC < 3 $\log_{10}$ CFU/cm^2 by 11.9%.	Used on small areas or hot spots only prior to chilling. Effectivness depends on employee skill, equipment maintenance, exposure time, application temperature.		✓	✓	Brashears and Chaves, 2017; Dorsa, Cutter and Siragusa, 1996; Hochreutener *et al.*, 2017; Moxley and Acuff, 2014; Zhilyaev *et al.*, 2017; Bacon *et al.*, 2002
Head and cheek meat	Head washing using water simulated plant conditions at pre-evisceration point using modified spray-wash cabinet; pre-evisceration wash (25 ± 20 °C) for 10 s at 3.2 kg/cm^2, followed by water (74 ± 2 °C) for 10 s at 0.7 kg/cm^2. Hot water applied for 26 s at 0.71 kg/cm^2 and at 74 ± 20 °C. Rate of hot water spray not measured. *E. coli* O157:H7 (no Stx) on beef cheeks from inoculated beef heads (n=140) was reduced by ≥ 1.5 $\log_{10}$ CFU/cm^2) after a pre-evisceration wash and a further 1.72 $\log_{10}$ CFU/cm^2 using hot water at 74 °C.	Targeted at cheek meat specifically; contamination can accumulate at head when vertical rail dressing; requires wash cabinet.	✓			Kal-chayanand *et al.*, 2008
Head washing using chemicals	Both lactic acid and FreshFx solutions were sprayed for 26 s at 1.75 kg/cm^2 and at 25 ± 2 °C and sprayed at the rate of 14 L/ min. Acidic electrolysed water (EOI) was sprayed for 26 s at 1.75 kg/cm^2 and 25 ± 2 °C; alkaline electrolysed water applied for 13 s at 1.75 kg/cm^2, followed by EO-I treatment for 13 s at 1.75 kg/cm^2. Ozonated water (OZI) applied for 26 s at 1.75 kg/cm^2 and 25 ± 2 °C; second treatment was a high-pressure water wash (HP; 0.2 kg/cm^2 at 25 °C) applied for 6 s, followed by OZI for 20 s at 1.75 kg/cm^2. Reductions ($\log_{10}$ CFU/cm^2) were achieved relative to a pre-evisceration wash using lactic acid (1.52), FreshFXTM (1.06) washes while reductions using electrolysed and ozonated water washes were not significantly different.	Targeted at cheek meat specifically; contamination can accumulate at head when vertical rail dressing; requires wash cabinet; GHP.	✓			Kal-chayanand *et al.*, 2008

TYPES EVIDENCE TO SUPPORT INTERVENTION	DETAILS OF RESEARCH STUDIES	OTHER CONSIDERATIONS	DEGREE OF SUPPORT			CITATIONS
			LOW	MED	HIGH	
3.4 Pre-chilling						
Washing using cold or ambient water	Inconsisten results form multiple studies. Spray beef carcasses 5.62 kg/cm^2, 32°C for 15 s with tap water (pH 7.34). Initial wash with water reduced STEC O157:H7 by more than 1.5 log$_{10}$ and reduced *Listeria* and *Clostridium* by 3 log$_{10}$. STEC O157:H7 and *Salmonella* Typhimurium were reduced 2.3 log$_{10}$. Apply carcass rinse of 1.5 L handwash (9 s at 69 kPa) and 5 L automated cabinet wash for 9 s. Meta-analysis of before and after effects conducted in commercial (large or small) slaughterhouses and pilot plants for 4 study consisting of 10 trials for prevalence and 3 studies comprising 9 trials for concentration estimates.	Redistribute contamination on carcass.	✓			Milios *et al.*, 2017; Dorsa, Cutter and Siragusa, 1996; Castillo *et al.*, 1999; Greig *et al.*, 2012; Gill, McGinnis and Badoni, 1996a, 1996b; McEvoy *et al.*, 1999; Yalçin *et al.*, 2004; Jericho *et al.*, 1995; Jericho *et al.*, 1996
Washing using hot water	Hot water pasteurization is defined as sheets of water applied to a carcass at temperatures greater than or equal to 85 °C for 8 s to 15 s. Results based on systematic literature review and meta-analysis. Effect size or intervention effectiveness was measured as raw log$_{10}$ reduction, least-squares means were calculated.	May generate condensate and aerosols; pressure of spray, health and safety issues for operators; colour changes.			✓	Castillo *et al.*, 2002; Dickson and Anderson, 1992; Gorman *et al.*, 1995; Smith and Graham, 1978; Smith, 1992; Greig *et al.*, 2012; Zhilyaev *et al.*, 2017; Huffman, 2002; Phetx-umphou, 2018.
Hot water and chilling	Carcass wash using water greater than 50°C followed by 24 h of chilling. Meta-analysis of before and after effects conducted in commercial (large or small) slaughterhouses and pilot plants for 1 study consisting of 2 trials. Odds ratio of NTS *E. coli* carcass contamination = 0.02.				✓	Greig *et al.*, 2012

TYPES EVIDENCE TO SUPPORT INTERVENTION	DETAILS OF RESEARCH STUDIES	OTHER CONSIDERATIONS	DEGREE OF SUPPORT			CITATIONS
			LOW	MED	HIGH	
Steam pasteurization	Steam pasteurization was defined as steam applied to a carcass at a temperature great than or equal to 82.2 °C for 6–11 s. Some observed reductions in aerobic plate counts of around 1.5 $\log_{10}$ cfu/cm^2 and the reduction of coliforms to below detectable levels, following a 6-8 s treatment. Use of a similar system gave consistent results and showed that the reduction was uniform over the surface of the carcass. Meta-analysis of before and after effects conducted in commercial (large or small) slaughterhouses and pilot plants for 4 study consisting of 14 trials. Plus 1 controlled trial with natural pathogen exposure. Odds ratio of NTS *E. coli* carcass contamination = 0.13. Controlled trial demonstrated a standardized mean difference of 0.39 $\log_{10}$	Surface greying of carcasses, but after 24 h chilling, the meat returned to acceptable colour.			✓	Nutsch *et al.*, 1997; Phebus, 1997; Greig *et al.*, 2012; Huffman, 2002
Organic acids – Lactic acid	In a laboratory study, lactic acid (2%) at 55°C was shown to reduce STEC O157:H7 on inoculated beef carcass tissue by 2.7 $\log_{10}$. Water gave a 1.6 $\log_{10}$ reduction 2.07 (95% CI 1.48, 2.65).	Concentration and type of acids used; regulatory requirements.	✓			Zhilyaev *et al.*, 2017; Ransom *et al.*, 2003; USDA/FSIS, 2021; EFSA, 2011; FDA, 2003; FDA, 2021a; MLA, no date
Organic acids - Acetic acid	STEC O157:H7 on inoculated beef carcass - 2% acetic acid reduced levels by 1.4 $\log_{10}$ in a laboratory study. Water gave a 1.6 $\log_{10}$ reduction. Using 2% acetic acid on beef brisket fat for 12 s immediately after being inoculated with faecal matter. STEC O157:H7 was reduced by 3.69 $\log_{10}$.		✓			Ransom *et al.*, 2003; Cabedo, Sofos and Smith, 1996; Zhilyaev *et al.*, 2017
Oxidizer-type antimicrobials	Peroxyacetic acid is approved by USDA FSIS for washing, rinsing, cooling, or otherwise processing fresh beef carcasses. Under laboratory conditions, researchers have achieved between 1-1.4 $\log_{10}$ reductions in STEC O157:H7 inoculated onto beef carcass tissue. In a commercial trial, the effect of a solution of 0.02% peroxyacetic acid on chilled beef quarters was investigated at two slaughter plants. The study found little effect on total bacteria or *E. coli* levels on meat from one of the plants, and no effect in the other plant.	Regulatory requirements.	✓			Ransom *et al.*, 2003; Gill and Badoni, 2004; FDA, 2021b

CONTROL MEASURES FOR SHIGA TOXIN-PRODUCING *ESCHERICHIA COLI* (STEC) ASSOCIATED WITH MEAT AND DAIRY PRODUCTS

TYPES EVIDENCE TO SUPPORT INTERVENTION	DETAILS OF RESEARCH STUDIES	OTHER CONSIDERATIONS	DEGREE OF SUPPORT			CITATIONS
			LOW	MED	HIGH	
Electrolysed (EO) water	Electrolysed (EO) water is produced by passing of electrical current through a dilute saltwater solution. One product of the reaction is sodium hydroxide (NaOH) and the other is hypochlorous acid, which has a low pH, contains active chlorine, and has a strong oxidation-reduction potential similar to that of ozone.		✓			Wheeler, Kalchayanand and Bosilevac, 2014
Oxidizer-type antimicrobials – acidified sodium chlorite (ASC)	Some studies have demonstrated a 1.9–2.3 $\log_{10}$ reduction in Salmonella and STEC O157:H7 on beef carcass tissue using a wash or spray of sodium chlorite activated with citric acid. One laboratory trial showed up to 4.6 $\log_{10}$ reduction in STEC O157:H7 and Salmonella using a water wash followed by an acidified sodium chlorite spray. Other studies indicated limited success, and found that spray treatment with acidified sodium chlorite was not as effective at reducing STEC O157:H7 on beef flanks as spray treatments with hot water, lactic acid or peroxyacetic acid.	Method of activation and application and the contact time with the meat surface.	✓			Ransom et al., 2003; Castillo et al., 1999; Gill and Badoni, 2004; Kalchayanand et al., 2012
Ozone	STEC was reduced between 0.6-1.0 $\log_{10}$ on beef samples when exposed to 72 ppm of ozone. The results of that study indicated no difference in numbers of STEC O157:H7 and S. Typhimurium detected on the surfaces of a hot carcass after exposure to water wash containing 95 ppm ozone as compared to water alone.	Oxidation of fat and muscle pigments.	✓			Coll Cárdenas et al., 2011; Castillo et al., 2003
Oxidizer-type antimicrobials - Sodium hypochlorite (NaOCl)	Beef carcasses sprayed (4.22 kg/cm²; 4.2 L/min) NaOCl solution with 50, 100, 250, 500, and 800 ppm of chlorine at 28 °C. E. coli was reduced from 0.5 - 1.28 $\log_{10}$ CFU/cm² by these treatments, but the reduction is not significantly different from that of water.		✓			Cutter and Siragusa, 1995
Trisodium phosphate	Sprayed onto beef carcasses 5.62 kg/cm², 32°C for 15 s with 12% trisodium phosphate (pH 12.31). Initial wash with water reduced STEC O157:H7 by more than 2.5 $\log_{10}$ and reduced Listeria and Clostridium by 3 $\log_{10}$.		✓			Dorsa, Cutter and Siragusa, 1996

TYPES EVIDENCE TO SUPPORT INTERVENTION	DETAILS OF RESEARCH STUDIES	OTHER CONSIDERATIONS	DEGREE OF SUPPORT			CITATIONS
			LOW	MED	HIGH	
Combination of steam and lactic acid	Steam + lactic acid is defined as steam pasteurization (steam applied to a carcass at a temperature greater than or equal to 82.2°C for 6–11 s.) followed by a rinse of 2% lactic acid. STEC O157:H7 was reduced between 1 to 1.5 $\log_{10}$ with rinse of 1%, 3%, or 5% acetic, lactic, or citric acid; STEC O157:H7 and *Salmonella* Typhimurium were reduced 3.8 $\log_{10}$; STEC O157:H7 and *Salmonella* Typhimurium were reduced 4.5 $\log_{10}$; odds ratio of NTS *E. coli* carcass contamination = 0.01.		✓			Cutter and Siragusa, 1994; Castillo *et al.*, 1999; Greig *et al.*, 2012
Dry chill	Dry chill is defined as chilling following final carcass wash without the use of an acid or water spray chilling. Conventional chilling can reduce the microbial populations on carcasses by 0.3-0.7 $\log_{10}$, and can reduce *E. coli* counts by up to 2 $\log_{10}$ over 24-36 h. Some research shows APC loss and then recovery when simulated carcass conditions are used in a broth system. Meta-analysis of before and after effects conducted in commercial (large or small) slaughterhouses and pilot plants for 4 study consisting of 9 trials. Odds ratio of NTS *E. coli* carcass contamination = 0.17.		✓			Bacon *et al.*, 2000; Nortjé, and Naudé, 1981; Thomas *et al.*, 1977; McEvoy *et al.*, 2004; Gill, 1986; Chang *et al.*, 2003; Mellefont, Kocharunch-itt and Ross, 2015; Greig *et al.*, 2012; Gill and Bryant, 1997
Carcass spray chilling	Spray-chilling had only little effect on microbial populations when it is used. There was a higher likelihood of detecting *E. coli* after spray chilling.		✓			Greer and Dilts, 1988; Kinsella *et al.*, 2006; Greig *et al.*, 2012

TYPES EVIDENCE TO SUPPORT INTERVENTION	DETAILS OF RESEARCH STUDIES	OTHER CONSIDERATIONS	DEGREE OF SUPPORT			CITATIONS
			LOW	MED	HIGH	
Carcass chilling	In all experiments, the inactivating effects of oxidants were greatest on fat surfaces and much less effective on lean surfaces. ClO_2 at 15 ppm, caused higher $\log_{10}$ reductions in *E. coli* numbers (approximately 3 $\log_{10}$ reduction) when applied during spray chilling than when applied immediately prior to "normal" spray chilling (approximately 1 $\log_{10}$ reduction). Abattoir trial: spray chilling treatments (with water alone, peroxyacetic acid, PAA at 200 ppm or chlorine dioxide, ClO2 at 50 ppm); water alone was effective at the hindquarters (hind legs and bung), indicator bacteria substantially reduced. Antimicrobial, either PAA (200 ppm) or ClO2 (50 ppm) was added to the spray chill water, the reduction in indicator bacteria was enhanced at all carcass sites, especially hindquarters, NTS *E. coli* eliminated.	STEC might become more susceptible to oxidative damage when exposed to carcass chilling. (Chlorine dioxide, ClO_2 or peroxyacetic acid, PAA) on beef meat during a simulated spray chilling process (sprayed for 4 s every 15 min for 36 cycles) and/or when applied (sprayed for 144 s) prior to spray chilling with water.	✓			King *et al.*, 2016; Ko-charunchitt, 2020

Annex 3

Post-processing control strategies for STEC in beef

TYPES EVIDENCE TO SUPPORT INTERVENTION	DETAILS OF RESEARCH STUDIES	OTHER CONSIDERATIONS	DEGREE OF SUPPORT			CITATIONS
			LOW	MED	HIGH	
4.1 Physical Interventions						
Air-drying heat treatment	Treatment at 60 °C to 100 °C for 5 s to 600 s; beef cuts for catering; STEC O157:H7 reduction of 1.3 -6.1 $\log_{10}$ CFU/cm^2.	Impact on appearance and colour; selection of resistant subpopulation; laboratory based study.	✓	✓		McCann *et al.*, 2006
Condensing Steam	Treatment at 75 °C for 10 s at 38.6 Kpa; reduction of STEC O157:H7 by 1.5 $\log_{10}$ CFU cm^2 using meat slices.	Laboratory study.	✓			Logue, Sheridan and Harrington, 2005
Hot water	Water at 82 °C sprayed for 20 s; STEC O157:H7 reduction of 1.0 $\log_{10}$ CFU/100 cm^2 on sub-primals before mechanical tenderization-blade tenderization.	Can cause temperature change.	✓			Heller *et al.*, 2007
	Water at 82 °C, aerobically or anaerobically (559 mm/Hg vacuum) for 3 min; *Escherichia coli* (ATCC 11775; EC); hot water treatment of beef trimmings before grinding did not reduce any microorganism.	Laboratory study.	✓			Stivarius *et al.*, 2002
	Inoculated trimmings were exposed to hot water treatment, using about 23 L of water at 95 °C for 3 s (3 s was required for the surface to reach 82 °C). Beef trimmings from young or mature cattle were treated with hot water and challenged with STEC O157:H7 had a reduction of 0.9 $\log_{10}$ CFU/g (5.2-4.3 $\log_{10}$).	Laboratory study.	✓			Ellebracht *et al.*, 1999
Surface trimming	The external surface was trimmed away using a sterile knife. Subprimals with antimicrobial interventions before mechanical tenderization/blade tenderization had a STEC O157:H7 reduction of 1.1 $\log_{10}$ CFU/100 cm^2.	Loss of products.	✓			Heller *et al.*, 2007

TYPES EVIDENCE TO SUPPORT INTERVENTION	DETAILS OF RESEARCH STUDIES	OTHER CONSIDERATIONS	DEGREE OF SUPPORT			CITATIONS
			LOW	MED	HIGH	
Surface trimming	Partial-surface trimming, full-surface trimming; STEC O157:H7 inoculated on vacuum packaged sub-primals. High-inoculum level: reduction of 4.0 $\log_{10}$ CFU/cm^2 (from 4.8 ≤ 0.7). Low-inoculum level: reduction of 2.0 $\log_{10}$ CFU/cm^2 (from 2.9 ≤ 0.7).	Laboratory based.		✓		Jacob et al., 2011
Dry chilled ageing	All samples were suspended in a cold room (3 °C), with four defrost time periods of 20 min, an air velocity of 0.25 m/s, and a relative humidity of 80%. STEC O157:H7 on beef had a reduction of 4 $\log_{10}$ CFU/cm^2.	Laboratory based.	✓			
	Packaging aerobically, stored for 5 d at 7 °C. Beef cuts inoculated with STEC O157:H7; Reduction between 1.9-2.2 $\log_{10}$ (if previously vacuum packed at 12 °C).	Sensory changes not assessed.	✓			Ashton et al., 2006
	Blast Freezing: Bovine bulk manufactured cartons of beef; in frozen storage (minimum -18 °C) for a minimum period of 6 weeks. Seven STEC serogroups (O157, O26, O103, O111, O121, O145 and O45) showed a reduction by 70% of positive lots.	Plant based study.		✓		Koh, 2020
High pressure processing (HPP)	Vacuum packaged sub-primals, 1 to 2 °C, 120 days. Neither carcass nor intervention treatment had any significant (P > 0.05), beneficial effect on the microbiological quality of sub-primal cuts. HPP treated ground beef. Single-cycle. 400 MPa 12 °C for a 1- to 20-min cycle. Multiple-cycle four 1-min cycles at 400 MPa and 12 °C and three 5-min cycles; 4-7 °C, 450 MPa and 15 min.	Laboratory data.		✓		Kenney et al., 1995; Morales et al., 2008; Hsu et al., 2015
Irradiation	Irradiation (eBeam) 1 KGy. Sub-primals prior to mechanical tenderization inoculated with STEC and STEC O157:H7 - 4 $\log_{10}$ reduction. Irradiation (eBeam) on ground beef of 1 KGy. Irradiation (Gamma) of 2.5 KGy and of 0.5 or 2 KGy on trim, prior to grind.	Laboratory data, costly equipment; no effect on sensory quality below 5 KGy.	✓			Kundu et al., 2014; Arthur et al., 2005; Arthur et al., 2005; De la Paz Xavier et al., 2014; Cap et al., 2020
Irradiation and packaging	Ground ground beef packaged at 4°C; < 2 KGy. MAP treatment of ground beef meat balls. Irradiation (gamma) at 1.5 KGy in MAP (3% O_2+ 50% CO_2+ 47% N_2) or aerobic packages. MAP/vacuum packaging of Ground beef patties. Irradiation (eBeam) at 0.5, 1, or 1.5 KGy in MAP (99.6% CO_2, 0.4% CO) or vacuum.	Lab data, costly equipment.		✓		Sommers et al., 2015; Gunes et al., 2011; Kudra et al., 2011

TYPES EVIDENCE TO SUPPORT INTERVENTION	DETAILS OF RESEARCH STUDIES	OTHER CONSIDERATIONS	DEGREE OF SUPPORT			CITATIONS
			LOW	MED	HIGH	
Irradiation and organic acid	Trim treated with radiation (eBeam)/lactic acid; Lactic acid (LA) – 5% (55 °C), 1 KGy in aerobic or vacuum packages (4 °C). Trim treated with radiation (gamma)/lactic acid/caprylic acid: Lactic acid (LA) 0.5% (50 °C); Caprylic acid (CA) – 0.04% (50 °C); 3. 0.5 - 2 KGy (12 °C).	Laboratory data; costly equipment.		✓		Li et al., 2015; Cap et al., 2020

4.2 Chemical Interventions

			LOW	MED	HIGH	
Organic acids	Lactic acid treatment of sub-primals, trim and cheek meat; various studies using STEC O157:H7 strains and/or NTS E. coli. Range of reductions achieved - STEC O157:H7, 0.2-2.8 $\log_{10}$; NTS E. coli 0.2-3.4 $\log_{10}$.	Review of 20 studies (18 involved artificial inoculation, 1 controlled trial and 1 before and after trial).		✓		Antic, 2018
	Hydroxypropanoic acid: Thin slices from sub-primals and primal cores inoculated with a 4- strain cocktail of STEC O157:H7; incubated for 1h-14 d at 4 °C. STEC O157:H7 reduction: 1 h -0.67 $\log_{10}$; 1 d - 0.89 $\log_{10}$; 7 d -1.47 $\log_{10}$.	Pilot plant tenderization; No sensory evaluation: Residual antimicrobial activity questions.	✓			Muriana et al., 2019
	Lactic acid dip; Beef trim for ground beef. Two 4 strain cocktails - STEC O157:H7 and non O157 STEC. Treatment used 4.4% lactic acid (ambient temperature) dip and spray treatments – 5 s dip or 13 s spray - samples tested after 1 h and after 20 h vacuum packed at 4°C then later ground to produce ground beef.	Sensorial quality could change depending on the exposure time.	✓			Wolf et al., 2012
Other chemical treatments	Sulfuric acid-sodium sulfate blend (SSS): Mixtures of STEC O157:H7 (5 strains), non-O157 STEC (12 strains). 2 spray treatment levels on pre-rigor beef resulted in 0.6–1.5 $\log_{10}$ CFU/cm^2 reduction.	Laboratory study, safety implications.	✓			Scott-Bullard et al., 2017

TYPES EVIDENCE TO SUPPORT INTERVENTION	DETAILS OF RESEARCH STUDIES	OTHER CONSIDERATIONS	DEGREE OF SUPPORT			CITATIONS
			LOW	MED	HIGH	
Other chemical treatments	Hypobromous acid, neutral acidified sodium chlorite, and two citric acid-based compounds used in treatment of ground beef: Cocktail 1 was STEC O26, O103, O111, O145, and O157; Cocktail 2 STEC O45, O121, O157 and Salmonella seeded at high cfu (1.5×10^7) and low cfu (1.5×10^4) inoculum as a spray for 15s and then quantified after 10 min and 48 h at 4 °C with 48 h storage. The treatments resulted in 0.7- 2.3 $\log_{10}$ reductions of STEC. Reductions of 2 $\log_{10}$ or more were achieved for O26, O103 and O145 but only with the high inoculum and after chill storage following citric acid or acidified sodium chlorite. One of the citric acid products tested gave a reduction of 2 $\log_{10}$ for O103 following treatment. In low inoculum study none of the treatments eliminated the 7 STEC strains tested.	Laboratory study	✓			Kal-chayanand et al., 2015
	Disodium metasilicate: Thin slices from sub-primals and primal cores, 1 h - 14 d at 4 °C. 4 strain cocktail of STEC O157:H7. Surface spray 6%, 18-20 °C; pH 13. STEC O157:H7 reduction: 1 h - 1.06 $\log_{10}$; 1 d - 2.07 $\log_{10}$; 7 d - 3.61 $\log_{10}$.	Pilot plant tenderization; no sensory evaluation; antimicrobial activity concerns.		✓		Muriana et al., 2019
	Thin slices from sub-primals and primal cores inoculated with a 4 strain of STEC O157:H7 cocktail, incubated for 1 h-14 d at 4 °C. Treated with Lauric arginate and Peroxyacetic acid. Surface spray 5000 ppm (LA) and 220 ppm (PA). STEC O157:H7 reduction: 1 h - 1.16 $\log_{10}$; 1 d - 1.95 $\log_{10}$; 7 d - 2.18 $\log_{10}$.	Pilot plant tenderization; no sensory evaluation; antimicrobial activity concerns.		✓		Muriana et al., 2019
Ozone	Exposed to ozone (72 ppm) in a continuous ozonation chamber at 0 °C and 4 °C during 3 h and 24 h incubation. Impact on E. coli (STEC not specified) was 0.6-1.0 $\log_{10}$ reduction.	Impact on colour and lipid oxidation.	✓			Coll Cárdenas et al., 2011
	Minced beef, 500-5000 ppm dry ozone gas.	Laboratory data, costly equipment, impractical to work with these concentrations.	✓			McMillin and Michel, 2000
Lactoferricin B	Ground beef - Lactoferricin B (100µg/g) added and stored at 4 °C and 10 °C for 3 days to control a 5 strain STEC O157:H7 cocktail inoculated at 10^7 cfu/ml.	Laboratory data, no sensory assessment of impact on TVC.	✓			Venkita-narayanan, Zhao and Doyle, 1999
Essential oils	Thyme oil treatment of minced beef - 0.6% thyme oil then stored at 4 °C and 10 °C for up to 12 days; 2 strains of STEC O157:H7.	Laboratory data; sensory changes were acceptable.	✓			Solomakos et al., 2008

TYPES EVIDENCE TO SUPPORT INTERVENTION	DETAILS OF RESEARCH STUDIES	OTHER CONSIDERATIONS	DEGREE OF SUPPORT			CITATIONS
			LOW	MED	HIGH	
4.3 Biological Interventions						
Bacterio-phages	Primals treated with phage, 7 $\log_{10}$ PFU/4 cm^2, STEC O157:H7 had a 2.7 $\log_{10}$ reduction (37 °C). Primals treated with a 6-phage cocktail - ~9 $\log_{10}$ PFU/mL. STEC and STEC O157:H7 seeded 9 $\log_{10}$ cfu/ml had a reduction of 0.5 - 1 $\log_{10}$ (4 °C) and 3-3.8 $\log_{10}$ (37 °C) in 48 h. Sub-primals treated with a phage cocktail - 10 $\log_{10}$ PFU. STEC and STEC O157:H7 reduced by 0.77 $\log_{10}$ (3 h), 1.15 $\log_{10}$ (6 h).	Laboratory data; not suitable for food processing; emergence of phage insensitive mutants; only tested on small pieces of meats.	✓			Hudson *et al.*, 2013; Tomat *et al.*, 2013a, 2018
	During packaging, 3 phage cocktails.	Laboratory data, takes too long.	✓			Hong, Pan and Ebner, 2014
Lactic acid bacteria (LAB)	Lactic acid bacteria (LAB) with vacuum packaging; LAB inoculum - $\log_{10}$ 8.7 cfu/ml, 4 LAB strains - inoculum $\log_{10}$ 7 cfu/ml. LAB inoculum with ageing strips of meat – $\log_{10}$ 8.7 cfu/ml; STEC and STEC O157:H7. 0.4 $\log_{10}$/cm^2 reduction (4 °C), 14-28 days.	Laboratory data, takes too long; no effect on sensory quality.	✓			Smith *et al.*, 2005a; Kirsch *et al.*, 2017
Colicins	Sub-primals of pork prior to tenderization, 3 mg colicin M + 1 mg colicin E7/kg. STEC O157:H7 seeded at 5 $\log_{10}$; reductions of 2.3 $\log_{10}$ in 1 h and 2.7 $\log_{10}$ in 1 d (10 °C).	Laboratory data; economically made in tobacco tissue; limited data for pork.	✓			Schulz *et al.*, 2015
Combinations of post-processing interventions						
Combinations of Steam/Vacuum	Meat slices treated with condensing steam at 75 °C, following vacuum packaging and stored in air or under vacuum at 0 °C; STEC O157:H7 reduction of 1.5 $\log_{10}$ CFU/g.	Laboratory data		✓		Logue, Sheridan and Harrington, 2005
Combinations of MAP or vacuum packaging/ lactic acid	Steaks treated with 10% lactic acid; STEC O157:H7 reduction of 2 $\log_{10}$ CFU/g.	Some loss of colour	✓	✓		Salim *et al.*, 2017
Combinations of lactic acid and hot water	Raw beef treated with 4% lactic acid, 80 °C, 20 s; STEC O157:H7 reduction of 3 $\log_{10}$ CFU/g.	Cost	✓	✓		Buncic *et al.*, 2014
Combinations of vacuum packaging, ambient water and organic acids	Treatment of meat inoculated with STEC O157:H7, non-O157 STEC (O26, O103, O111, and O145) treated with ambient water, 200 ppm Hypobromous acid, 200 ppm peroxyacetic acid, and 5% lactic acid. Spray and stored vacuum packed at 4 °C for 14 d. STEC O157:H7 reduction of 1.6-2.1 $\log_{10}$ CFU/50 cm^2; non-O157 STEC reduction was smaller at 0.4 -0.3 $\log_{10}$ CFU/50 cm^2.	Laboratory study	✓			Liao *et al.*, 2015

TYPES EVIDENCE TO SUPPORT INTERVENTION	DETAILS OF RESEARCH STUDIES	OTHER CONSIDERATIONS	DEGREE OF SUPPORT			CITATIONS
			LOW	MED	HIGH	
Combinations of Lauric arginate, water or other chemicals	Beef trim - Spray treatment with lauric acid (LA) 5% or followed by 0.4% cetylpyridinum chloride (CC), 4% sodium metasilicate (SM), 0.02% peroxyacetic acid (PA), 10% trisodium phosphate (TSP) or sterile water (SW) prior to grinding. Incubated at 4 °C for up to 3 d under simulated retail conditions. Inoculated with cocktail including 1 STEC O157:H7 and 6 non-O157 STEC strains.	Laboratory testing storage simulated retail conditions, no impact on colour of ground beef.	✓	✓		Dias-Morse et al., 2014
Combinations of essential oil and colicin (nisin)	Minced beef treated with thyme oil (essential oils) and colicin (nisin). 4 °C and 10 °C for up to 12 d; 0.6% thyme oil; 2 strains 4°C and 10 °C for up to 12 d; 0.6% thyme + nisin (500 IU/ml).	Laboratory data; sensory changes were acceptable.	✓			Solomakos et al., 2008
Combinations of irradiation (eBeam) and lactic acid	Trim treated with radiation (eBeam)/lactic acid; Lactic acid (LA) – 5% (55 °C); 1 KGy in aerobic or vacuum packages (4 °C).	Laboratory data; costly equipment.		✓		Li et al., 2015
Combinations of irradiation (eBeam) and packaging	Combination of radiation (eBeam)/MAP/vacuum packaging of ground beef patties; 0.5, 1, or 1.5 KGy in MAP (99.6% CO_2, 0.4% CO) or vacuum.	Laboratory data, costly equipment		✓		Kudra et al., 2011
Combinations of irradiation (gamma) and organic acids	Trim treated with radiation (gamma)/lactic acid/caprylic acid; Lactic acid (LA) 0.5% (50 °C); Caprylic acid (CA) – 0.04% (50 °C); 0.5 - 2 KGy (12 °C).	Laboratory data; costly equipment.		✓		Cap et al., 2020
Combinations of irradiation (gamma) and MAP	Radiation (gamma)/MAP treatment of ground beef meat balls; 1.5 KGy in MAP (3% O_2+ 50% CO_2+ 47% N_2) or aerobic packages.	Laboratory data, costly equipment.		✓		Gunes et al., 2011
Combinations of HPP and vacuum packaging	Frozen ground beef vacuum-packaged, pressure-treated at 400 MPa for10 min at -5 °C or 20 °C and stored at -20 °C or 4 °C for 5–30 d; four, 60 s cycles, 400 MPa, 17 °C.	Laboratory based		✓		Black et al., 2010; Jiang et al., 2015
High Pressure Processing (HPP) combinations	HPP and freezing of ground beef 25 °C, 400 MPa at five pressure cycles of 3 min.	GAP; additional studies to assess further sensory changes in ground beef; cost.		✓		Zhou, Karwe and Matthews, 2016

Annex 4

Processing and post-processing control strategies for STEC in raw milk and raw milk cheese

TYPES EVIDENCE TO SUPPORT INTERVENTION	DETAILS OF RESEARCH STUDIES	OTHER CONSIDERATIONS	DEGREE OF SUPPORT			CITATIONS
			LOW	MED	HIGH	
5.1 Raw milk processing						
Bactofugation	Milk heated to 55 °C to 60 °C, *Enterobacteriaceae* 72% removal; double bactofugation at 54.4 °C, removed 95% of *E. coli*.	Cost, volume that can be processed at one time. Laboratory-based data. No STEC data.	✓			Faccia *et al.*, 2013
Microfiltration	1.4 µm ceramic filter; skim milk only and at 50 °C. Effectively removed 3-5 $\log_{10}$ CFU/ml to be non-detectable.	Cost, volume that can be processed at one time. No *E. coli* or STEC data. Laboratory-based data. Some data on cold (6 °C) microfiltration.	✓			Elwell and Barbano, 2006
High Pressure Processing (HPP)	400 MPa at 50 °C for 15 m; 5 $\log_{10}$ CFU/mL reduction of STEC O157:H7 in 15 m in UHT milk. Earlier study observed significant strain to strain variations. 400 MPa at 21 °C to 31 °C for 50 m; 6 $\log_{10}$ CFU/mL reduction of *E. coli* in 30 m in human milk but 8 $\log_{10}$ CFU/mL after 10 m in peptone solution. *E. coli* ATCC 25922 (Stx negative) was inactivated by 8 $\log_{10}$ after 10 m in peptone solution and by 6 $\log_{10}$ after 30 m in human milk.	Temperature used may not be acceptable for definition of "raw milk". Laboratory-based data. Cost, volume that can be processed from bovine or ruminant milk. Limited data for NTS *E. coli* only. Laboratory based data.	✓	✓		Patterson and Kilpatric, 1998; Viazis, Farkas and Jaykus, 2008
Irradiation (cold pasteurization)	eBeam at 1 to 2 kGy; 1 kGy eliminated coliform; 2 kGy - 4 $\log_{10}$ CFU/mL reduction in aerobic count; STEC O157:H7 D10 value - 0.062 kGy.	Off-flavors in dairy products in laboratory-based studies.			✓	Ward, Kerth and Pillai, 2017
Bacteriophage	6 phage cocktail,- 9 $\log_{10}$ PFU/ml used in inoculated challenge studies with NTS *E. coli* and STEC O157:H7. *E. coli* and STEC O157:H7 - seeded at 4 $\log_{10}$ CFU/mL. 2 $\log_{10}$ CFU/mL reduction in 1 d at 4 °C.	Takes too long for practical applications; sensory attributes not affected. Laboratory based data.	✓			Tomat *et al.*, 2018

TYPES EVIDENCE TO SUPPORT INTERVENTION	DETAILS OF RESEARCH STUDIES	OTHER CONSIDERATIONS	DEGREE OF SUPPORT			CITATIONS
			LOW	MED	HIGH	
7.3 Laboratory testing for STEC detection across the dairy processing chain *7.3.1.Raw milk*						
STEC and stx gene in raw milk and milk filters	In study over 1-year STEC O157:H7 recovered from 2% of milk filters but not from raw milk. Stx gene detected in 37% of milk filters and 7% of raw milk. Filters are more effective for detection of pathogens than reliance on sampling from the bulk tank. It has been used on farm, so is technically feasible.	Filter type, duration, location flow rate of milk. Milk filters can be optimized for bacterial detection.				Jaakkonen *et al.*, 2019
Sampling plans	Large-scale milk collection: universal sampling plan (bulk tank samples); small-scale milk collection: periodic, random or composite sampling all have roles but depending on local conditions. In Germany, STEC are included in the testing of bulk tank milk from cattle, sheep and goats.roles but depending on local conditions. Suggested indicator criteria: APC - n=5, c=2 m=2x104, M=5x104; Enterobacteriaceae- n=5, c=1, m=10, M=102; S. aureus - n=5, c=2, m=10, M=102 In developed and certain developing countries, testing raw milk filters can offer a better sampling point that bulk tank sampling. In Italy, researchers identified the presence of STEC O157:H7, O45, O103, O121 and O145 in bulk tank milk and raw milk filter samples collected at commercial dairy plants.	Limitations of bulk tank and plant sampling; sampling and testing plans vary by country and region; under unfavourable hygiene and temperature conditions in the supply chain; microbial limits should be adjusted as the situation improves. Most testing plans target *E. coli* and other indicator organisms. Only a few countries (e.g. Germany) test for STEC; other do so based on risk assessment.				FDA/ USPHS, 2017; Draaiyer *et al.*, 2009; ICMSF, 2011; FAO/ WHO, 2018; Albonico *et al.*, 2017
5.2 Raw milk cheese processing *5.2.1 Milk fermentation*						
Milk fermentation	Autochthonous strains of LAB: The bacteriocins produced by isolates showed antimicrobial activity towards different spoilage and pathogenic microorganisms.	Laboratory data for *Listeria monocytogenes*, *Staphylococcus aureus*, *Clostridium tyrobutyricum* and *Brochothrix thermosphacta*.	✓			Dal Bello *et al.*, 2010
5.2.2 Protective cultures						
Protective cultures	Addition of a consortium of H. alvei, Lactobacillus plantarum and Lactococcus lactis - Average of 2.8 log$_{10}$ CFU/g reduction of STEC O26:H11 in uncooked pressed cheeses made from different raw milk batches.	Variation in reduction depending on the milk microbiota (positive/ negative interactions between the microbiota, consortium and STEC); in pilot plant challenge studies.	✓			Fretin *et al.*, 2020

TYPES EVIDENCE TO SUPPORT INTERVENTION	DETAILS OF RESEARCH STUDIES	OTHER CONSIDERATIONS	DEGREE OF SUPPORT			CITATIONS
			LOW	MED	HIGH	
5.2.3 Bacteriophage						
Bacteriophage	Phage cocktail - Inactivated *E. coli* and STEC O157:H7 in 8 h. Without phage treatment, NTS *E. coli* grew to 4-6 $\log_{10}$ cfu/g in 6 h.	Starter culture - not affected; under laboratory-based conditions.	✓			Tomat *et al.*, 2013b
5.2.4 Acidification, salting and cooking						
Lactic acid - dry salting on surface (for lactic goat cheeses)	Temperature/time 24 °C/24 h; decrease of pH (4.21 at day 2) versus variety of STEC. No increase (except 1 $\log_{10}$ CFU/g for STEC O26 during the first hours). Reduction of STEC O26, O103, O145 and O157 to levels below enumeration limit (i.e. <10 CFU/g) (but still detectable by enrichment).	Effective in pilot plant challenge studies. Acidification (pH < 4.3) is necessary for efficacy.		✓		Miszczycha *et al.*, 2013
Rennet - pressing - brining (for uncooked pressed cheese with short ripening)	Temperature/time 34 °C/30 m; decrease of pH (5.3 at day 1). Increase of 3.3 to 5 $\log_{10}$ CFU/g ue to growth and entrapment.	Growth variations between serotypes; acidification must be below pH 5 for efficacy. Studies performed in pilot plants.	✓			Miszczycha *et al.*, 2013
Acidification, salting and cooking	Sharp increase of 2.2 to 3 $\log_{10}$ CFU/g, after 6 h, for cheeses inoculated with STEC O26:H11 at 0.05 and 0.5 CFU/mL respectively.	pH must be < 5.0 for efficacy; in pilot plant challenge studies.	✓			Fretin *et al.*, 2020
Rennet - pressing - dry salting of curd (for uncooked pressed cheese with long ripening)	Temperature/time 32 °C/45 m decrease of pH (5.19 at day 1). Increase of 2 $\log_{10}$ CFU/g (STEC O157:H7), 4 $\log_{10}$ CFU/g (STEC O26) due to growth and entrapment. The pH must be < 5.0 for efficacy.	Growth variation between serotypes (faster for STEC O26) during pilot plant challenge studies.	✓			Miszczycha *et al.*, 2013
Rennet - dry salting on surface (for blue type cheese)	Temperature/time 32.5 °C/1 h; decrease of pH (4.91 at day 3). Increase of 1 $\log_{10}$ CFU/g (STEC O157:H7), 2 $\log_{10}$ CFU/g (STEC O103), 3 $\log_{10}$ CFU/g (STEC O26) due to growth and entrapment. pH must be < 4.91 for efficacy.	Growth variation between serotypes (faster for STEC O26) in pilot plant challenge studies.	✓			Miszczycha *et al.*, 2013
Rennet - brining (for white mold cheese)	Temperature/time 32.5 °C/1 h; decrease of pH (4.91 at day 3). Increase of 2 $\log_{10}$ CFU/g (STEC O157:H7) to 3 $\log_{10}$ CFU/g (STEC O26, O103, and O145), cdue to growth and entrapment. pH must be < 5.0 for efficacy.	Growth variation between serotypes.	✓			Miszczycha *et al.*, 2016

CONTROL MEASURES FOR SHIGA TOXIN-PRODUCING *ESCHERICHIA COLI* (STEC) ASSOCIATED WITH MEAT AND DAIRY PRODUCTS

TYPES EVIDENCE TO SUPPORT INTERVENTION	DETAILS OF RESEARCH STUDIES	OTHER CONSIDERATIONS	DEGREE OF SUPPORT			CITATIONS
			LOW	MED	HIGH	
Rennet - curd cooking - brining - pressing (for cooked pressed cheese)	Temperature/time 54 °C/35 m; decrease of pH (5.38 at day 1). No increase. Reduction of STEC O26, O103, O145 and O157 to below enumeration limit at 1.75 h. Cooking step (> 53 °C) needed for efficacy.	Pilot plant challenge test.		✓		Miszczycha et al., 2013
Curd cooking	Temperature/time, 53 °C/20 m. > 4.5 $\log_{10}$ CFU/g reduction of heat-sensitive NTS E. coli. Cooking step (>53°C) needed for efficacy. 0.5 $\log_{10}$ CFU/g reduction of thermotolerant NTS E. coli.	2 NTS E. coli behaving similar to STEC in pilot plant challenge study.		✓		Peng et al., 2013b
Acidification - brine -curd cooking	Temperature/time/acidification/brine 53 °C to 40 °C/2 h and further steps/8 kg. > 4 $\log_{10}$ CFU/g reduction of thermotolerant NTS E. coli. Cooking step (> 53 °C) needed for efficacy.	2 NTS E. coli behaving similar to STEC in pilot plant challenge study.		✓		Peng et al., 2013b
5.2.5 Ripening and ageing						
Lactic goat cheese	At 4 °C for 20 d then at 8 °C for 35 d. Increase of pH to 5.26 at day 25 (constant till day 60). The a_w decreased in the core from 0.994 (day 2) to 0.967 (day 45). Further reduction for 4/8 strains (not detectable by enrichment at day 60). Ripening: efficient to prevent growth after acidification (but not for elimination).	Effective in pilot plant challenge studies.		✓		Miszczycha et al., 2013
Cooked pressed cheese	At 9 °C - 10 °C for 4 months; pH 5.38 stable till day 30 then increase to 5.82 at day 120. Core: a_w 0.975 at day 120. Ripening: efficient to prevent growth after cooking (but not for elimination). In the core: not detected except for 1/4 replicate (STEC O157:H7); In the rind, detected by enrichment for 3/4 replicates.	Variation in cell levels between core/rind. Effective in pilot plant challenge studies.		✓		Miszczycha et al., 2013
Uncooked pressed cheese with short ripening	Uncooked pressed cheese with short ripening at 12 °C for 12 d - 20 d then 4 °C for 12-20 d then 8 °C for 8 d; increase of pH to 5.80 at day 40. The a_w in the core remained constant (0.974). Ripening: not efficient for reduction. $a_w >$ 0.95 (minimum a_w). No reduction (STEC levels remained constant till day 40).	Effective in pilot plant challenge studies.	✓			Miszczycha et al., 2013

TYPES EVIDENCE TO SUPPORT INTERVENTION	DETAILS OF RESEARCH STUDIES	OTHER CONSIDERATIONS	DEGREE OF SUPPORT			CITATIONS
			LOW	MED	HIGH	
Uncooked pressed cheese with long ripening	Uncooked pressed cheese with long ripening at 9 °C -10 °C for 7 months. Small increase of pH to 5.52 in the core (constant till day 240). pH higher in the rind (7.34-7.64). In the core, the a_w decreased slowly to reach 0.943 at day 240. In the rind, a_w decreased to 0.922 at day 240. Reduction (>2 log_{10} CFU/g) from day 60–240. STEC O157:H7 and STEC O26 below enumeration limit (except for STEC O26, 3 log_{10} CFU/g core). Ripening: efficient for reduction (if duration > 60 d), but not for elimination	Variation between serotypes (STEC O26 more persistent inside the core). Effective in pilot plant challenge studies.		✓		Miszczycha et al., 2013
Blue type cheese	At 11 °C (16 d), -2 °C (125 d), 4 °C (30 d), 12 °C (60 d). Increase of pH to 6.69 at day 25, then decrease from day 100 to day 240 (pH 5.5). The a_w gradually decreased to reach 0.898 at day 240. Reduction depending on time and serotypes: STEC O157:H7 and STEC O103 detectable at day 60 but not at day 240; STEC O26 detected at day 240 only after enrichment. Ripening: efficient (if duration >60 d) for reduction, but not for elimination. Combination of negative temperature, very low a_w 0.898 (< 0.95 minimum aw), acidic pH and impact of *Penicilium roqueforti*.	Variation between serotypes (STEC O26 more persistent). Effective in pilot plant challenge studies.		✓		Miszczycha et al., 2013
White mold cheese	At 13 °C (14 d), 4°C (14 d), 8 °C (28 d). Increase of pH to 5.7-6.1 (core) or 6.5 (rind) at day 56. The a_w gradually decreased but remained > 0.96 at day 56. Efficient for STEC O157:H7, not for non-O157 STEC. a_w > 0.95 (minimum a_w); high pH (rind). Reduction depending on serotypes: 1-2 log_{10} CFU/g (STEC O26, O103 and O145); 2-3 log_{10} CFU/g (STEC O157:H7; detected only after enrichment). Higher STEC level in rind compared to core.	Variation between serotypes (STEC O157:H7 less persistent) and in core/ rind in pilot plant challenge studies.	✓			Miszczycha et al., 2016
Soft cheese	Ripening at 11 °C for 2 weeks, efficient for STEC O157:H7, less efficient for non-O157 STEC. The survival probability after 2 weeks ripening was 1%, 34%, 37%, and 27%, respectively, for STEC O157:H7, O103:H2, O26:H11, and O145:H28.	Variation between serotypes (STEC O157:H7 less persistent) in pilot plant challenge studies.	✓			Perrin et al., 2015

TYPES EVIDENCE TO SUPPORT INTERVENTION	DETAILS OF RESEARCH STUDIES	OTHER CONSIDERATIONS	DEGREE OF SUPPORT			CITATIONS
			LOW	MED	HIGH	
Hard cheese	Cooking at 53 °C, 11 kg, up to 16 weeks ripening. Positive for thermotolerant *E. coli* in 2/8 samples by enrichment after 4 weeks. Positive for heat-sensitive *E. coli* in 1/12 samples by enrichment after 16 weeks.	Temperature gradient from rind to core during first day, samples taken close to rind; effective in pilot plant challenge studies.	✓			Peng *et al.*, 2013b
Semi-hard cheese	Cooking at 46 °C, 7.5 kg, up to 16 weeks ripening. Approx. 0.4 $\log_{10}$ CFU/g reduction/week in cheese core. Approx. 0.2 $\log_{10}$ CFU/g reduction/week in cheese rind low level of *E. coli* in raw milk (approx. 100 CFU/mL): < LOQ in cheese core after 8 weeks, but detectable by enrichment after 16 weeks. In rind, *E. coli* was quantifiable until end of ripening.	2 NTS *E. coli* behaving similar to STEC; effective in pilot plant challenge studies.	✓			
	Cooking at 46 °C, 250g, up to 16 weeks ripening. 0.25-0.79 $\log_{10}$ CFU/g reduction/week (depending on strain). 1 STEC showed fastest reduction: not possible to enrich after 16 weeks, others had no significant difference.	3 non-O157-STEC and 2 NTS *E. coli*, monitored separately in pilot plant challenge studies.	✓			Peng *et al.*, 2013b
	Cooking at 40 °C, 250g, up to 16 weeks ripening. 0.23-0.58 $\log_{10}$ CFU/g reduction/week (depending on strain from NTS *E. coli* and 3 non O157 STEC). 1 STEC showed fastest reduction: not possible to enrich after 16 weeks, others had no significant reduction.	3 non-O157-STEC and 2 NTS *E. coli*, monitored separately in pilot plant challenge studies.	✓			

5.2.6 *Cheese size*

TYPES EVIDENCE TO SUPPORT INTERVENTION	DETAILS OF RESEARCH STUDIES	OTHER CONSIDERATIONS	DEGREE OF SUPPORT			CITATIONS
			LOW	MED	HIGH	
Cheese Size	Temperature gradient in hard and extra hard cheeses after cooking at first 24 h. For information: heat map of temperature gradient.	Sampling of rind and core differed. Smaller size cheese differed from larger cheeses and impacted STEC survival in laboratory versus production scale cheese making.	✓			Ercolini *et al.*, 2005

TYPES EVIDENCE TO SUPPORT INTERVENTION	DETAILS OF RESEARCH STUDIES	OTHER CONSIDERATIONS	DEGREE OF SUPPORT			CITATIONS
			LOW	MED	HIGH	
5.3 Raw milk cheese post-processing						
Packaging	Modified atmosphere packaging (MAP) can retard growth of pathogens in a hard cheese made of raw sheep milk, but a higher decrease in numbers of *L. monocytogenes* and *S. aureus* were observed than for STEC O157:H7. Active packaging is used the food industry mainly to extend shelf-life, but also to prevent growth of pathogens.	Active packaging technologies applications in cheese are still very limited and further studies are needed to explore the potential and efficiency of these new technologies.	✓			Solomakos *et al.*, 2019; Yildirim *et al.*, 2018; Speranza *et al.*, 2020; Al-Moghazy, Mahmoud and Nada, 2020
Irradiation (eBeam)	0.2 - 2 kGy, Brie and Camembert - 1.27–2.59 kGy, controlled growth.	Tested with *L. monocytogenes*; no sensory defect in laboratory-based study		✓		Velasco *et al.*, 2015
Bacteriophage	3-phage cocktail; multiplicity of infection - 1; at 24 °C - STEC O157:H7 seeded at 7 $\log_{10}$ CFU/g. No reduction in STEC in cheese slices at room temperature.	Emergence of phage resistant STEC O157:H7 (and potentially other STEC). Studies performed in cheese slices under laboratory conditions.	✓			Hong, Pan and Ebner, 2014
7.3 Laboratory testing for STEC detection across the dairy processing chain *7.3.1.Raw milk*						
Sampling plans	Systematic selection of quality raw milk with *E. coli* counts < 50 CFU/mL for raw milk cheese manufacturing was predicted to reduce HUS risk. Various criteria could reduce the risk of HUS by 25% to 89% with probabilities of noncompliance between 3% and 79%. Increasing analysis sample size of end products from 25 g to 100 g for STEC detection was predicted to reduce HUS risk. In cheeses made from pasteurized milk The ICMSF (2011) recommended *E. coli* limits established under a 3-class sampling plan where n = 5, c = 3, m = 10 and M = 100. Raw milk cheese is tested for *S. aureus* only, consistent with EU recommended sampling criteria.	There are general recommendations for microbiological sampling programs to assure hygiene and safety, but none are specific to STEC. It is notable that for EU microbiological criteria for cheese, no limits were established for STEC in raw milk cheese. Differences by type of cheese.				ICMSF, 2011; Perrin *et al.*, 2015; Anses opinion, 2018; Donnelly, 2018

References for annexes 1-4

Ahmad, A., Nagaraja, T. G. & Zurek, L. 2007. Transmission of *Escherichia coli* O157:H7 to cattle by house flies. *Preventative Veterinary Medicine,* 80: 74–81. doi: 10.1016/j. prevetmed.2007.01.006

Albonico, F., Gusmara, C., Gugliotta, T., Loiacono, M., Mortarino, M. & Zecconi, A. 2017. A new integrated approach to analyze bulk tank milk and raw milk filters for the presence of the *E. coli* serogroups frequently associated with VTEC status. *Research in Veterinary Science,* 115: 401–406. doi: 10.1016/j.rvsc.2017.07.019

Allison, H. E. 2007. Stx-phages: drivers and mediators of the evolution of STEC and STEC-like pathogens. *Future Microbiology,* 2: 165–174. doi: 10.2217/17460913.2.2.165

Al-Moghazy, M., Mahmoud, M. & Nada, A. A. 2020. Fabrication of cellulose-based adhesive composite as an active packaging material to extend the shelf life of cheese. *International Journal of Biologic Macromolecules,* 160: 264–275. doi: 10.1016/j. ijbiomac.2020.05.217

ANSES. 2018. Avis de l'Anses relatif au protocole de reprise de la commercialisation de reblochons proposé par l'entreprise Chabert. 2018. www.anses.fr/fr/system/files/ BIORISK2018SA0164.pdf

Antic, D. 2018. A critical literature review to assess the significance of intervention methods to reduce the microbiological load on beef through primary production. University of Liverpool. Food Standards Agency Project FS301044. https://www. food.gov.uk/print/pdf/node/4506

Antic, D., Blagojevic, B., Ducic, M., Mitrovic, R., Nastasijevic, I. & Buncic, S. 2010a. Treatment of cattle hides with Shellac-in-ethanol solution to reduce bacterial transferability--a preliminary study. *Meat Science,* 85: 77–81. doi: 10.1016/j. meatsci.2009.12.007

Antic, D., Blagojevic, B., Ducic, M., Nastasijevic, I., Mitrovic, R. & Buncic, S. 2010b. Distribution of microflora on cattle hides and its transmission to meat via direct contact. *Food Control,* 21: 1025–1029. doi: 10.1016/j.foodcont.2009.12.022

Antic, D., Blagojevic, B. & Buncic, S. 2011. Treatment of cattle hides with shellac solution to reduce hide-to-beef microbial transfer. *Meat Science,* 88: 498–502. doi: 10.1016/j.meatsci.2011.01.034

Arthur, T. M., Barkocy-Gallagher, G. A., Rivera-Betancourt, M. & Koohmaraie, M. 2002. Prevalence and characterization of Non-O157 Shiga toxin-producing *Escherichia coli* on carcasses in commercial beef cattle processing plants. *Applied and Environmental Microbiology,* 68(10): 4847–4852. doi: 10.1128/AEM.68.10.4847-4852.2002

Arthur, T. M., Wheeler, T. L., Shackelford, S. D., Bosilevac, J. M., Nou, X. & Koomaraie, M. 2005. Effects of low-dose, low-penetration electron beam irradiation of chilled beef carcass surface cuts on *Escherichia coli* O157:H7 and meat quality. *Journal of Food Protection,* 68: 666–672. doi: 10.4315/0362-028X-68.4.666

Arthur, T. M., Bosilevac, J. M., Brichta-Harhay, D. M., Guerini, M. N., Kalchayanan, N., Shackleford, S. D., Wheeler, T. L. & Koohmaraie, M. 2007a. Transportation and lairage environment effects on prevalence, numbers, and diversity of *Escherichia coli* O157:H7 on hides and carcasses of beef cattle at processing. *Journal of Food Protection*, 70: 280–286. doi: 10.4315/0362-028x-70.2.280

Arthur, T. M., Bosilevac, J. M., Brichta-Harhay, D.M., Kalchayanand, N., Shackelford, S. D., Wheeler, T. L. & Koohmaraie, M. 2007b. Effects of a minimal hide wash cabinet on the levels and prevalence of *Escherichia coli* O157:H7 and *Salmonella* on the hides of beef cattle at slaughter. *Journal of Food Protection*, 70: 1076–1079. doi: 10.4315/0362-028x-70.5.1076

Arthur, T. M., Bosilevac, J. M., Brichta-Harhay, D. M., Kalchayanand, N., King, D. A., Shackelford, S. D., Wheeler, T. L. & Koohmaraie, M. 2008. Source tracking of *Escherichia coli* O157:H7 and *Salmonella* contamination in the lairage environment at commercial U.S. beef processing plants and identification of an effective intervention. *Journal of Food Protection*, 71: 1752–1760. doi: 10.4315/0362-028x-71.9.1752

Arthur, T. M., Kalchayanand, N., Agga, G. E., Wheeler, T. L. & Koohmaraie, M. 2017. Evaluation of bacteriophage application to cattle in lairage at beef processing plants to reduce *Escherichia coli* O157:H7 prevalence on hides and carcasses. *Foodborne Pathogens and Disease*, 14: 17–22. doi: 10.1089/fpd.2016.2189

Ashton, L. V., Geornaras, I., Stopforth, J. D., Skandamis, P. N., Belk, K. E., Scanga, J. A., Smith, G. C. & Sofos, J. N. 2006. Fate of inoculated *Escherichia coli* O157:H7, cultured under different conditions, on fresh and decontaminated beef transitioned from vacuum to aerobic packaging. *Journal of Food Protection*, 69(6): 1273-1279. doi: 10.4315/0362-028x-69.6.1273

Avery, S. M., Small, A., Reid, C.-A. & Buncic, S. 2002. Pulsed-field gel electrophoresis characterization of Shiga toxin-producing *Escherichia coli* O157 from hides of cattle at slaughter. *Journal of Food Protection*, 65(7): 1172–1176. doi: 10.4315/0362-028X-65.7.1172

Ayscue, P., Lanzas, C., Ivanek, R. & Gröhn, Y. T. 2009. Modeling on-farm *Escherichia coli* O157:H7 population dynamics. *Foodborne Pathogens and Disease*, 6(4): 461–470. doi: 10.1089/fpd.2008.0235

Bach, S. J., McAllister, T. A., Baah, J., Yanke, L. J., Veira, D. M., Gannon, V. P. J. & Holley, R. A. 2002. Persistence of *Escherichia coli* O157:H7 in barley silage: effect of a bacterial inoculant. *Journal of Applied Microbiology*, 93(2): 288–294. doi: 10.1046/j.1365-2672.2002.01695.x

Bach, S. J., Selinger, J., Stanford, K. & McAllister, T. A. 2005a. Effect of supplementing corn- or barley-based feedlot diets with canola oil on faecal shedding of *Escherichia coli* O157:H7 by steers. *Journal of Applied Microbiology*, 98: 464–75. doi: 10.1111/j.1365-2672.2004.02465.x

Bach, S. J., Stanford, K. & McAllister, T. A. 2005b. Survival of *Escherichia coli* O157:H7 in feces from corn- and barley-fed steers. *FEMS Microbiology Letters*, 252: 25–33. doi: 10.1016/j.femsle.2005.08.030

Bacon, R. T., Belk, K. E., Sofos, J. N., Clayton, R. P., Reagan, J. O. & Smith, G. C. 2000. Microbial populations on animal hides and beef carcasses at different stages of slaughter in plants employing multiple-sequential interventions for decontamination. *Journal of Food Protection*, 63(8): 1080–1086. doi: 10.4315/0362-028X-63.8.1080

Bacon, R. T., Sofos, J. N., Belk, K. E. & Smith, G. C. 2002. Application of a commercial steam vacuum unit to reduce inoculated *Salmonella* on chilled fresh beef adipose tissue. *Dairy, Food and Environmental Sanitation*, 22: 184–190.

Baird, B. E., Lucia, L. M., Acuff, G. R., Harris, K. B. & Savell, J. W. 2006. Beef hide antimicrobial interventions as a means of reducing bacterial contamination. *Meat Science*, 73(2): 245–248. doi: 10.1016/j.meatsci.2005.11.023

Baumberger, C., Guarín, J. F. & Ruegg, P. L. 2016. Effect of two different premilking teat sanitation routines on reduction of bacterial counts on teat skin of cows on commercial dairy farms. *Journal of Dairy Science*, 99: 2915–2929. doi: 10.3168/jds.2015-10003

Beauvais, W., Gart, E. V., Bean, M., Blanco, A., Wilsey, J., McWhinney, K., Bryan, L., Krath, M., Yang, C. Y., Alvarez, D.M., Paudyal, S., Bryan, K., Stewart, S., Cook, P. W., Lahodny, G., Baumgarten, K, Gautam, R., Nightingale, K., Lawhon, S. D., Pinedo P. & Ivane, R. 2018. The prevalence of *Escherichia coli* O157:H7 fecal shedding in feedlot pens is affected by the water-to-cattle ratio: A randomized controlled trial. *PLOS ONE*. 13: e0192149. doi: 10.1371/journal.pone.0192149

Berg, J. L., McAllister, T. A., Bach, S. J., Stillborn, R. P., Hancock D. D. & LeJeune, J. T. 2004. *Escherichia coli* O157:H7 excretion by commercial feedlot cattle fed either barley- or corn-based finishing diets. *Journal of Food Protection*, 67: 666–671. doi: 10.4315/0362-028x-67.4.666

Bergholz, T. M. & Whittam, T. S. 2007. Variation in acid resistance among enterohaemorrhagic *Escherichia coli* in a simulated gastric environment. *Journal of Applied Microbiology*, 102(2): 352–362. doi: 10.1111/j.1365-2672.2006.03099.x

Berry, E. D. & Wells, J. E. 2010. *Escherichia coli* O157:H7: Recent advances in research on occurrence, transmission, and control in cattle and the production environment. *Advances in Food and Nutrition Research*, pp. 67–117. Elsevier. doi: 10.1016/S1043-4526(10)60004-6

Berry, E. D., Wells, J. E., Varel, V. H., Hales K. E. & Kalchayanand, N. 2017. Persistence of *Escherichia coli* O157:H7 and total *Escherichia coli* in feces and feedlot surface manure from cattle fed diets with and without corn or sorghum wet distillers grains with solubles. *Journal of Food Protection*, 80: 1317–1327. doi: 10.4315/0362-028X.JFP-17-018

Besser, T. E., Schmidt, C. E., Shah, D. H. & Shringi, S. 2014. "Preharvest" food safety for *Escherichia coli* O157 and other pathogenic Shiga toxin-producing strains. *Microbiology Spectrum*, 2: 21–28. doi: 10.1128/microbiolspec.EHEC-0021-2013

Black, E. P., Hirneisen, K. A., Hoover, D. G. & Kniel, K. E. 2010. Fate of *Escherichia coli* O157:H7 in ground beef following high-pressure processing and freezing. *Journal of Applied Microbiology*, 108: 1352–1360. doi: 10.1111/j.1365-2672.2009.04532.x

Blagojevic, B., Antic, D., Ducic, M. & Buncic, S. 2012. Visual cleanliness scores of cattle at slaughter and microbial loads on the hides and the carcasses. *Veterinary Record*, 170: 563. doi: 10.1136/vr.100477

Blaustein, R. A., Pachepsky, Y. A., Shelton, D. R. & Hill, R. L. 2015. Release and removal of microorganisms from land-deposited animal waste and animal manures: a review of data and models. *Journal of Environmental Quality*, 44: 1338–1354. doi: 10.2134/jeq2015.02.0077

Bosilevac, J. M., Arthur, T. M., Wheeler, T. L., Shackelford, S. D., Rossman, M., Reagan, J. O. & Koohmaraie, M. 2004. Prevalence of *Escherichia coli* O157 and levels of aerobic bacteria and *Enterobacteriaceae* are reduced when hides are washed and treated with cetylpyridinium chloride at a commercial beef processing plant. *Journal of Food Protection*, 67(4): 646–650. doi: 10.4315/0362-028X-67.4.646

Bosilevac, J. M., Shackelford, S. D., Brichta, D. M. & Koohmaraie, M. 2005a. Efficacy of ozonated and electrolyzed oxidative waters to decontaminate hides of cattle before slaughter. *Journal of Food Protection*, 68: 1393–1398. doi: 10.4315/0362-028x-68.7.1393

Bosilevac, J. M., Nou, X., Osborn, M. S., Allen, D. M., Koohmaraie, M. 2005b. Development and evaluation of an on-line hide decontamination procedure for use in a commercial beef processing plant. *Journal of Food Protection*, 68(2): 265–272. doi: 10.4315/0362-028x-68.2.265

Bosilevac, J. M., Nou, X., Barkocy-Gallagher, G. A., Arthur, T. M. & Koohmaraie, M. 2006. Treatments using hot water instead of lactic acid reduce levels of aerobic bacteria and *Enterobacteriaceae* and reduce the prevalence of *Escherichia coli* O157:H7 on preevisceration beef carcasses. *Journal of Food Protection*, 69: 1808–1813. doi: 10.4315/0362-028x-69.8.1808

Bosilevac, J. M., Wang, R., Luedtke, B. E., Hinkley, S., Wheeler, T. L. & Koohmaraie, M. 2017. Characterization of enterohemorrhagic *Escherichia coli* on veal hides and carcasses. *Journal of Food Protection*, 80(1): 136–145. doi: 10.4315/0362-028X.JFP-16-247

Braden, K. W., Blanton Jr, J. R., Allen, V. G., Pond, K. R. & Miller, M. F. 2004. *Ascophyllum nodosum* supplementation: A preharvest intervention for reducing *Escherichia coli* O157:H7 and *Salmonella* spp. in feedlot steers. *Journal of Food Protection*, 67: 1824–1828. doi: 10.4315/0362-028x-67.9.1824

Brashears, M. M. & Chaves, B. D. 2017. The diversity of beef safety: A global reason to strengthen our current systems. *Meat Science*, 132: 59–71. doi: 10.1016/j.meatsci.2017.03.015

Brown, M., ed. 2000. HACCP in the Meat Industry. 1st ed. Southampton: Woodhead Publishing. ISBN: 9781855736443

Brown-Brandl, T. M., Berry, E. D., Wells, J. E., Arthur, T. M. & Nienaber, J. A. 2009. Impacts of individual animal response to heat and handling stresses on *Escherichia coli* and *E. coli* O157:H7 fecal shedding by feedlot cattle. *Foodborne Pathogens and Disease*, 6: 855–864. doi: 10.1089/fpd.2008.0222.

Buchko, S. J., Holley, R. A., Olson, W. O., Gannon, V. P. J. & Veira, D. M. 2000. The effect of fasting and diet on fecal shedding of *Escherichia coli* O157:H7 by cattle. *Canadian Journal of Animal Science*, 80(4): 741–744. doi: 10.4141/A00-025

Buncic, S., Nychas, G. J., Lee, M. R. F., Koutsoumanis, K., Hébraud, M., Desvaux, M., Chorianopoulos, N., Bolton, D., Blagojevic, B. & Antic, D. 2014. Microbial pathogen control in the beef chain: Recent research advances. *Meat Science*, 97(3): 288–297. doi: 10.1016/j.meatsci.2013.04.040.

Byrne, C. M., Bolton, D. J., Sheridan, J. J., McDowell, D. A. & Blair, I. S. 2000. The effects of preslaughter washing on the reduction of *Escherichia coli* O157:H7 transfer from cattle hides to carcasses during slaughter. *Letters in Applied Microbiology*, 30(2): 142–145. doi: 10.1046/j.1472-765x.2000.00689.x

Cabedo, L., Sofos, J. N. & Smith, G. C. 1996. Removal of bacteria from beef tissue by spray washing after different times of exposure to fecal material. *Journal of Food Protection*, 59 (12): 1284–1287. doi: 10.4315/0362-028X-59.12.1284

Callaway, T. R., Anderson, R. C., Genovese, K. J., Poole, T. L., Anderson, T. J., Byrd, J. A., Kubena, L. F. & Nisbet, D. J. 2002. Sodium chlorate supplementation reduces *E. coli* O157:H7 populations in cattle. *Journal of Animal Science,* 80(6): 1683–1689. doi: 10.2527/2002.8061683x

Callaway, T. R., Edrington, T. S., Anderson, R. C., Genovese, K. J., Poole, T. L., Elder, R. O., Byrd, J. A., Bischoff, K. M. & Nisbet, D. J. 2003. *Escherichia coli* O157:H7 populations in sheep can be reduced by chlorate supplementation. *Journal of Food Protection*, 66: 194–199. doi: 10.4315/0362-028x-66.2.194

Callaway, T. R., Stahl, C. H., Edrington, T. S., Genovese, K. J., Lincoln, L. M., Anderson, R. C., Lonergan, S. M., Poole, T. L., Harvey, R. B. & Nisbet, D. J. 2004. Colicin concentrations inhibit growth of *Escherichia coli* O157:H7 in vitro. *Journal of Food Protection*, 67: 2603–2607. doi: 10.4315/0362-028x-67.11.2603

Callaway, T. R., Carr, M. A., Edrington, T. S., Anderson, R. C. & Nisbet, D. J. 2009. Diet, *Escherichia coli* O157:H7, and cattle: A review after 10 years. *Current Issues in Molecular Biology*, 11: 67–79. https://www.caister.com/cimb/v/v11/67.pdf

Callaway, T. R., Carroll, J. A., Arthington, J. D., Edrington, T. S., Anderson, R. C., Rossman, M. L., Carr, M. A., Genovese, K. J., Ricke, S. C., Crandall, P. & Nisbet, D. J. 2011. *Escherichia coli* O157:H7 populations in ruminants can be reduced by orange peel product feeding. *Journal of Food Protection*, 74: 1917–1921. doi: 10.4315/0362-028X.JFP-11-234

Callaway, T. R., Edrington, T. S., Loneragan, G. H., Carr, M. A. & Nisbet, D. J. 2013. Shiga toxin-producing *Escherichia coli* (STEC) ecology in cattle and management based options for reducing fecal shedding. *Agriculture Food and Analytical Bacteriology*, 3: 39–69. https://pubag.nal.usda.gov/download/57234/pdf

Callaway, T. R., Edrington, T. S. & Nisbet, D. J. 2014. Isolation of *Escherichia coli* O157:H7 and *Salmonella* from migratory brown-headed cowbirds (*Molothrus ater*), common grackles (*Quiscalus quiscula*), and cattle egrets (*Bubulcus ibis*). *Foodborne Pathogens and Disease*, 11(10): 791–794. doi: 10.1089/fpd.2014.1800

Cap, M., Cingolani, C., Lires, C., Mozgovoj, M., Soteras, T., Sucari, A., Gentiluomo, J., Descalzo, A., Grigioni, G., Signorini, M., Horak, C., Vaudagna, S. & Leotta, G. 2020. Combination of organic acids and low-dose gamma irradiation as antimicrobial treatment to inactivate Shiga toxin-producing *Escherichia coli* inoculated in beef trimmings: lack of benefits in relation to single treatments. *PLOS ONE*, 15(3):e0230812. doi: 10.1371/journal.pone.0230812

Carlson, B. A., Ruby, J., Smith, G. C., Sofos, J. N., Bellinger, G. R., Warren-Serna, W., Centrella, B., Bowling, R. A. & Belk, K. E. 2008. Comparison of antimicrobial efficacy of multiple beef hide decontamination strategies to reduce levels of *Escherichia coli* O157:H7 and *Salmonella*. *Journal of Food Protection*, 71: 2223–2227. doi: 10.4315/0362-028x-71.11.2223

Castillo, A., Dickson, J. S., Clayton, R. P., Lucia, L. M. & Acuff, G. R. 1998a. Chemical dehairing of bovine skin to reduce pathogenic bacteria and bacteria of fecal origin. *Journal of Food Protection*, 61: 623–625. doi: 10.4315/0362-028x-61.5.623

Castillo, A., Lucia, L. M., Goodson, K. J., Savell, J. W. & Acuff, G. R. 1998b. Comparison of water wash, trimming, and combined hot water and lactic acid treatments for reducing bacteria of fecal origin on beef carcasses. *Journal of Food Protection*, 61: 823–828. doi: 10.4315/0362-028x-61.7.823

Castillo, A., Lucia, L. M., Kemp, G. K. & Acuff, G. R. 1999. Reduction of *Escherichia coli* O157:H7 and *Salmonella* Typhimurium on beef carcass surfaces using acidified sodium chlorite. *Journal of Food Protection*, 62 (6): 580–584. doi: 10.4315/0362-028X-62.6.580

Castillo, A., Hardin, M. D., Acuff, G. R. & Dickson, J. S. 2002. Reduction of microbial contaminants on carcasses. In V.K. Juneja & J. N. Sofos, eds. *Control of foodborne microorganisms*, pp. 351–381. New York, CRC Press. doi: 10.1201/b16945–14

Cernicchiaro, N., Pearl, D. L., McEwen, S. A., Harpster, L., Homan, H. J., Linz, G. M. & LeJeune, J. T. 2012. Association of wild bird density and farm management factors with the prevalence of *E. coli* O157 in dairy herds in Ohio (2007-2009). *Zoonoses and Public Health*, 59: 320–329. doi: 10.1111/j.1863-2378.2012.01457.x

Cernicchiaro, N., Renter, D. G., Cull, C. A., Paddock, Z. D., Shi, X. & Nagaraja, T. G. 2014. Fecal shedding of non-O157 serogroups of Shiga toxin-producing *Escherichia coli* in feedlot cattle vaccinated with an *Escherichia coli* O157:H7 SRP vaccine or fed a *Lactobacillus*-based direct-fed microbial. *Journal of Food Protection*, 77(5): 732–737. doi: 10.4315/0362-028X.JFP-13-358

Cernicchiaro, N., Oliveira, A. R. S., Hoehn, A., Noll, L.W., Shridhar, P. B., Nagaraja, T. G., Ives, S. E., Renter, D. G. & Sanderson, M. W. 2020. Associations between season, processing plant, and hide cleanliness scores with prevalence and concentration of major Shiga toxin-producing *Escherichia coli* on beef cattle hides. *Foodborne Pathogens and Disease,* 17: 611–619. doi: 10.1089/fpd.2019.2778

Chang, V. P., Mills, E. W. & Cutter, C. N. 2003. Reduction of bacteria on pork carcasses associated with chilling method. *Journal of Food Protection,* 66(6): 1019–1024. doi: 10.4315/0362-028X-66.6.1019

Čížek, A., Alexa, P., Literák, I., Hamřík, J., Novák, P. & Smola, J. 1999. Shiga toxin-producing *Escherichia coli* O157 in feedlot cattle and Norwegian rats from a large-scale farm. *Letters in Applied Microbiology*, 28(6): 435–439. doi: 10.1046/j.1365-2672.1999.00549.x

Coffey, B., Rivas, L., Duffy, G., Coffey, A., Ross, R. P. & McAuliffe, O. 2011. Assessment of *Escherichia coli* O157:H7-specific bacteriophages e11/2 and e4/1c in model broth and hide environments. *International Journal of Food Microbiology*, 147: 188–194. doi: 10.1016/j.ijfoodmicro.2011.04.001

Colavecchio, A., Cadieux, B., Lo, A. & Goodridge, L. D. 2017. Bacteriophages contribute to the spread of antibiotic resistance genes among foodborne pathogens of the *Enterobacteriaceae* family - a review. *Frontiers in Microbiology*, 8: 1108. doi: 10.3389/fmicb.2017.01108

Coll Cárdenas, F., Andrés, S., Giannuzzi, L. & Zaritzky, N. 2011. Antimicrobial action and effects on beef quality attributes of a gaseous ozone treatment at refrigeration temperatures. *Food Control,* 22: 1442–1447. doi: 10.1016/j.foodcont.2011.03.006

Cook, K. L., Bolster, C. H., Ayers, K. A. & Reynolds, D. N. 2011. *Escherichia coli* diversity in livestock manures and agriculturally impacted stream waters. *Current Microbiology*, 63(5): 439–449. doi: 10.1007/s00284-011-0002-6

Cull, C. A., Renter, D. G., Dewsbury, D. M., Noll, L. W., Shridhar, P. B., Ives, S. E., Nagaraja, T. G. & Cernicchiaro, N. 2017. Feedlot- and pen-level prevalence of enterohemorrhagic *Escherichia coli* in feces of commercial feedlot cattle in two major U.S. cattle feeding areas. *Foodborne Pathogens and Disease*, 14(6): 309–317. doi: 10.1089/fpd.2016.2227

Cutter, C. N. & Siragusa, G. R. 1994. Efficacy of organic acids against *Escherichia coli* O157:H7 attached to beef carcass tissue using a pilot scale model carcass washer. *Journal of Food Protection,* 57(2): 97–103. doi: 10.4315/0362-028X-57.2.97

Cutter, C. N. & Siragusa, G. R. 1995. Application of chlorine to reduce populations of *Escherichia coli* on beef. *Journal of Food Safety,* 15: 67-75. doi: 10.1111/j.1745-4565.1995.tb00121.x

Dal Bello, B., Rantsiou, K., Bellio, A., Zeppa, G., Ambrosoli, R., Civera, T. & Cocolin, L. 2010. Microbial ecology of artisanal products from North West of Italy and antimicrobial activity of the autochthonous populations. *LWT - Food Science and Technology,* 43: 1151–1159. doi: 10.1016/j.lwt.2010.03.008

Dawson, D. E., Keung, J. H., Napoles, M. G., Vella, M. R., Chen, S., Sanderson, M. W. & Lanzas, C. 2018. Investigating behavioral drivers of seasonal Shiga-Toxigenic *Escherichia coli* (STEC) patterns in grazing cattle using an agent-based model. *PLOS ONE,* 13(10): e0205418. doi: 10.1371/journal.pone.0205418

De la Paz Xavier, M., Dauber, C., Mussio, P., Delgado, E., Maquieira, A., Soria, A., Curuchet, A., Márquez, R., Méndeza, C. & López, T. 2014. Use of mild irradiation doses to control pathogenic bacteria on meat trimmings for production of patties aiming at provoking minimal changes in quality attributes. *Meat Science,* 98: 383–391. doi: 10.1016/j.meatsci.2014.06.037

Depenbusch, B. E., Nagaraja, T. G., Sargeant, J. M., Drouillard, J. S., Loe, E. R. & Corrigan, M. E. 2008. Influence of processed grains on fecal pH, starch concentration, and shedding of *Escherichia coli* O157 in feedlot cattle. *Journal of Animal Science,* 86: 632–639. doi: 10.2527/jas.2007-0057

Dewell, G. A., Simpson, C. A., Dewell, R. D., Hyatt, D. R., Belk, K. E., Scanga, J. A., Morley, P. S., Grandin, T., Smith, G. C., Dargatz, D. A., Wagner, B. A. & Salman, M. D. 2008. Impact of transportation and lairage on hide contamination with *Escherichia coli* O157 in finished beef cattle. *Journal of Food Protection,* 71: 1114–1118. doi: 10.4315/0362-028x-71.6.1114

Dewsbury, D. M. A. 2015. Epidemiology of Shiga toxin-producing *Escherichia coli* in the bovine reservoir: seasonal prevalence and geographic distribution. *Department of Diagnostic Medicine and Pathobiology, Kansas State University.* (Master Thesis). Cited 30 August 2022. https://krex.k-state.edu/dspace/handle/2097/19127

Dias-Morse, P., Pohlman, F. W., Williams, J. & Brown, A. H. 2014. Single or multiple decontamination interventions involving lauric arginate on beef trimmings to enhance microbial safety of ground beef. *The Professional Animal Scientist,* 30(5): 477–484. doi: 10.15232/pas.2014-01334

Dickson, J. S. & Anderson, M. E. 1992. Microbiological decontamination of food animal carcasses by washing and sanitizing systems: A review. *Journal of Food Protection,* 55(2): 133–140. doi: 10.4315/0362-028X-55.2.133

Diez-Gonzalez, F., Callaway, T. R., Kizoulis, M. G. & Russell, J. B. 1998. Grain feeding and the dissemination of acid-resistant *Escherichia coli* from cattle. *Science,* 281: 1666–1668. doi: 10.1126/science.281.5383.1666

Dodd, C. C., Sanderson, M. W., Sargeant, J. M., Nagaraja, T. G., Oberst, R. D., Smith, R. A. & Griffin, D. D. 2003. Prevalence of *Escherichia coli* O157 in cattle feeds in midwestern feedlots. *Applied and Environmental Microbiology,* 69(9): 5243–5247. doi: 10.1128/AEM.69.9.5243-5247.2003

Donnelly C., 2018. Review of controls for pathogen risks in Scottish artisan cheeses made from unpasteurised milk. www.foodstandards.gov.scot/downloads/ FSS_2017_015_-_Control_of_pathogens_in_unpasteurised_milk_cheese_-_ Final_report_v_4.7_-_20th_November_2018_.pdf

Dorsa, W. J., Cutter, C. N. & Siragusa, G. R. 1996. Effectiveness of a steam-vacuum sanitizer for reducing *Escherichia coli* O157:H7 inoculated to beef carcass surface tissue. *Letters in Applied Microbiology,* 23: 61–63. doi: 10.1111/j.1472-765x.1996. tb00029.x

Draaiyer, J., Dugdill, B., Bennett, A. & Mounsey, J. 2009. *Resource book-A practical guide to assist milk producer groups.* Rome, FAO. 86pp. www.fao.org/3/y3548e/ y3548e.pdf

Duffy, G. 2003. Verocytoxigenic *Escherichia coli* in animal faeces, manures and slurries. *Journal of Applied Microbiology,* 94 Suppl: 94S–103S. doi: 10.1046/j.1365-2672.94. s1.11.x

Edrington, T. S., Callaway, T. R., Anderson, R. C., Genovese, K. J., Jung, Y. S., Elder, R. O., Bischoff, K. M. & Nisbet, D. J. 2003. Reduction of *E. coli* O157:H7 populations in sheep by supplementation of an experimental sodium chlorate product. *Small Ruminant Research,* 49: 173–181. doi: 10.1016/S0921-4488(03)00099-3

Edrington, T. S., Schultz, C. L., Genovese, K. J., Callaway, T. R, Looper, M. L., Bischoff, K. M., McReynolds, J. L., Anderson, R. C. & Nisbet, D. J. 2004. Examination of heat stress and stage of lactation (early versus late) on fecal shedding of *E. coli* O157:H7 and *Salmonella* in dairy cattle. *Foodborne Pathogens and Disease,* 1: 114–119. doi: 10.1089/153531404323143639

Edrington, T. S., Carter, B. H., Friend, T. H., Hagevoort, G. R., Poole, T. L., Callaway, T. R., Anderson, R. C. & Nisbet, D. J. 2009a. Influence of sprinklers, used to alleviate heat stress, on faecal shedding of *E. coli* O157:H7 and *Salmonella* and antimicrobial susceptibility of *Salmonella* and *Enterococcus* in lactating dairy cattle. *Letters in Applied Microbiology,* 48: 738–743. doi: 10.1111/j.1472-765X.2009.02603.x

Edrington, T. S., Farrow, R. L., Loneragan, G. H., Ives, S. E., Engler, M. J., Wagner, J. J., Corbin, M. J., Platter, W. J., Yates, D., Hutcheson, J. P., Zinn, R. A, Callaway, T. R., Anderson R. C. & Nisbet, D. J. 2009b. Influence of beta-agonists (Ractopamine HCl and Zilpaterol HCl) on fecal shedding of *Escherichia coli* O157:H7 in feedlot cattle. *Journal of Food Protection,* 72: 2587–2591. doi: 10.4315/0362-028X-72.12.2587

EFSA. 2011. Scientific opinion on the evaluation of the safety and efficacy of lactic acid for the removal of microbial surface contamination of beef carcasses, cuts and trimmings. *EFSA Journal*, 9: 2317–2343. doi: 10.2903/j.efsa.2011.2317

Ekong, P. S., Sanderson, M. W. & Cernicchiaro, N. 2015. Prevalence and concentration of *Escherichia coli* O157 in different seasons and cattle types processed in North America: A systematic review and meta-analysis of published research. *Preventative Veterinary Medicine*, 121: 74–85. doi: 10.1016/j.prevetmed.2015.06.019

Elder, R. O., Keen, J. E., Wittum, T. E., Callaway, T. R., Edrington, T. S., Anderson, R. C. & Nisbet, D. J. 2002. Intervention to reduce fecal shedding of enterohemorrhagic *Escherichia coli* O157:H7 in naturally infected cattle using neomycin sulfate. *Journal of Animal Science*, 80 (Suppl. 1): 15 (Abstr.). ISSN: 0021-8812.

Ellebracht, E. A., Castillo, A., Lucia, L. M., Miller, R. K. & Acuff, G. R. 1999. Reduction of pathogens using hot water and lactic acid on beef trimmings. *Journal of Food Science*, 64(6): 1094–1099. doi: 10.1111/j.1365-2621.1999.tb12289.x

Ellis-Iversen, J., Smith, R. P., Van Winden, S., Paiba, G. S., Watson, E., Snow, L. C. & Cook, A. J. 2008. Farm Practices to control *E. coli* O157 in young cattle--a randomised controlled trial. *Veterinary Research*, 39: 3–15. doi: 10.1051/vetres:2007041

Elmoslemany, A. M., Keefe, G. P., Dohoo, I. R., Wichtel, J. J., Stryhn, H. & Dingwell, R. T. 2010. The association between bulk tank milk analysis for raw milk quality and on-farm management practices. *Preventative Veterinary Medicine*, 95: 32–40. doi: 10.1016/j.prevetmed.2010.03.007

Elwell, M. W. & Barbano, D. M. 2006. Use of microfiltration to improve fluid milk quality. *Journal of Dairy Science*, 89: E10–E30. doi: 10.3168/jds.S0022-0302(06)72361-X

Ercolini, V., Fusco, G., Blaiotta, G., Sarghini, F. & Coppola, S. 2005. Response of *Escherichia coli* O157:H7, *Listeria monocytogenes*, *Salmonella* Typhimurium, and *Staphylococcus aureus* to the thermal stress occurring in model manufactures of grana padano cheese. *Journal of Dairy Science*, 88: 3818–3825. doi: 10.3168/jds.S0022-0302(05)73067-8

Eustace, I., Midgley, J., Giarrusso, C., Laurent, C., Jenson, I. & Sumner, J. 2007. An alternative process for cleaning knives used on meat slaughter floors. *International Journal of Food Microbiology*, 113: 23–27. doi: 10.1016/j.ijfoodmicro.2006.06.034

Faccia, M., Mastromatteo, M., Conte, A. & Del Nobile, M. A. 2013. Influence of the milk bactofugation and natural whey culture on the microbiological and physico-chemical characteristics of mozzarella cheese. *Journal of Food Processing and Technology*, 4: 1–7. doi: 10.4172/2157-7110.1000218

Fan, R., Shao, K., Yang, X., Bai, X., Fu, S., Sun, H., Xu, Y., Wang, H., Li, Q., Hu, B., Zhang, J. & Xiong, Y. 2019. High prevalence of non-O157 Shiga toxin-producing *Escherichia coli* in beef cattle detected by combining four selective agars. *BMC Microbiology*, 19(1): 213. doi: 10.1186/s12866-019-1582-8

FAO & WHO. 2005. Code of hygienic practice for meat (CAC/RCP 58-2005). Codex Alimentarius. Rome. FAO. www.fao.org/fao-who-codexalimentarius/sh-proxy/ru/?lnk=1&url=https%253A%252F%252Fworkspace.fao.org%252Fsites%252Fcodex%252FStandards%252FCXC%2B58-2005%252FCXP_058e.pdf

FAO & WHO. 2018. Shiga toxin-producing *Escherichia coli* (STEC) and food: attribution, characterization, and monitoring. *Microbiological Risk Assessment Series* No. 31. Rome. FAO. www.who.int/publications/i/item/9789241514279

FDA & USPHS (United States Food and Drug Administration & Public Health Service). 2017. Grade "A" pasteurized milk ordinance, including provisions from the grade "A" condensed and dry milk products and condensed and dry whey--supplement I to the grade "A" pasteurized milk ordinance. Public Health Service/Food and Drug Administration. 2017 Revision.

FDA. 2003. Code of Federal Regulations Title 21, Government Printing Office, USA. www.ecfr.gov/current/title-21

FDA. 2021a. Code of Federal Regulations Title 21(21 CFR 101.100 (a) (3)). Government Printing Office, USA. www.ecfr.gov/current/title-21/chapter-I/subchapter-B/part-101/subpart-G/section-101.100

FDA. 2021b. Code of Federal Regulations. Chemical oxygen generators. (21 CFR 173.368). Government Printing Office, USA. www.ecfr.gov/current/title-49/subtitle-B/chapter-I/subchapter-C/part-173/subpart-E/section-173.168

Feng, P., Lampel, K. A., Karch, H. & Whittam, T. S. 1998. Genotypic and phenotypic changes in the emergence of *Escherichia coli* O157:H7. *The Journal of Infectious Disease,* 177(6): 1750–1753. doi: 10.1086/517438

Fingermann, M., Avila, L., Belén De Marco, M., Vázquez, L., Di Biase, D.N., Müller, A. V., Lescano, M., Dokmetjian, J. C., Castillo, S. F & Pérez Quiñoy, J. L. 2018. OMV-based vaccine formulations against Shiga toxin producing *Escherichia coli* strains are both protective in mice and immunogenic in calves. *Human Vaccines and Immunotherapeutics*, 14(9): 2208–2213. doi: 10.1080/21645515.2018.1490381

Fox, J. T., Depenbusch, B. E., Drouillard, J. S. & Nagaraja, T. G. 2007. Dry-rolled or steam-flaked grain-based diets and fecal shedding of *Escherichia coli* O157 in feedlot cattle. *Journal of Animal Science,* 85: 1207–1212. doi: 10.2527/jas.2006-079

Frank, C., Kapfhammer, S., Werber, D., Stark, K. & Held, L. 2008. Cattle density and Shiga toxin-producing *Escherichia coli* infection in Germany: Increased risk for most but not all serogroups. *Vector-borne Zoonotic Disease*, 8: 635–643. doi: 10.1089/vbz.2007.0237

Free, A. L., Duoss, H. A., Bergeron, L. V., Shields-Menard, S. A., Ward, E., Callaway, T. R., Carroll, J. A., Schmidt, T. B. & Donaldson, J. R. 2012. Survival of O157:H7 and Non-O157 serogroups of *Escherichia coli* in bovine rumen fluid and bile salts. *Foodborne pathogens and disease*, 9(11): 1010-1014. doi: 10.1089/fpd.2012.1208

Fremaux, B., Raynaud, S., Beutin, L. & Rozand, C. V. 2006. Dissemination and persistence of Shiga toxin-producing *Escherichia coli* (STEC) strains on French dairy farms. *Veterinary Microbiology*, 117: 180–91. doi: 10.1016/j.vetmic.2006.04.030

Frétin, M., Chassard, C., Delbès, C., Lavigne, R., Rifa, E., Theil, S., Fernandez, B., Laforce, P. & Callon, C. 2020. Robustness and efficacy of an inhibitory consortium against *E. coli* O26:H11 in raw milk cheeses. *Food Control*, 115: 107282. doi: 10.1016/j.foodcont.2020.107282

Galton, D. M., Petersson, L. G. & Merrill, W. G. 1986. Effects of pre-milking udder preparation practices on bacterial counts in milk and on teats. *Journal of Dairy Science*, 69: 260–266. doi: 10.3168/jds.S0022-0302(86)80396-4

Garber, L., Wells, S., Schroeder-Tucker, L. & Ferris, K. 1999. Factors associated with fecal shedding of verotoxin-producing *Escherichia coli* O157 on dairy farms. *Journal of Food Protection*, 62: 307–12. doi: 10.4315/0362-028x-62.4.307

Gautam, R., Bani-Yaghoub, M., Neill, W. H., Döpfer, D., Kaspar, C. & Ivanek, R. 2011. Modeling the effect of seasonal variation in ambient temperature on the transmission dynamics of a pathogen with a free-living stage: example of *Escherichia coli* O157:H7 in a dairy herd. *Preventative Veterinary Medicine*, 102: 10–21. doi: 10.1016/j.prevetmed.2011.06.008

Gibson, H., Sinclair, L. A., Brizuela, C. M., Worton, H. L. & Protheroe, R. G. 2008. Effectiveness of selected premilking teat-cleaning regimes in reducing teat microbial load on commercial dairy farms. *Letters in Applied Microbiology*, 46: 295–300. doi: 10.1111/j.1472-765X.2007.02308.x

Gill, C. O. 1986. The microbiology of chilled meat storage. *Proceedings 24th Meat Industry Research Conference*, Hamilton, New Zealand. MIRINZ publication 852. https://meatupdate.csiro.au/new/Cold%20Treatments.pdf

Gill, C. O., McGinnis, J. C. & Badoni, M. 1996a. Use of total or *Escherichia coli* counts to assess the hygienic characteristics of a beef carcass dressing process. *International Journal of Food Microbiology*, 31(1-3): 181–196. doi: 10.1016/0168-1605(96)00982-8

Gill, C. O., Mcginnis, J. C. & Badoni, M. 1996b. Assessment of the hygienic characteristics of a beef carcass dressing process. *Journal of Food Protection*, 59(2): 136–140. doi: 10.4315/0362-028X-59.2.136

Gill, C. O. & Bryant, J. 1997. Assessment of the hygienic performance of two beef carcass cooling processes from product temperature history data or enumeration of bacteria on carcass surfaces. *Food Microbiology*, 14: 593-602. doi: 10.1006/fmic.1997.0120

Gill, C. O. & Badoni, M. 2004. Effects of peroxyacetic acid, acidified sodium chlorite or lactic acid solutions on the microflora of chilled beef carcasses. *International Journal of Food Microbiology*, 91: 43–50. doi: 10.1016/s0168-1605(03)00329-5

Gorman, B. M., Morgan, J. B., Sofos, J. N. & Smith, G. C. 1995. Microbiological and visual effects of trimming and/or spray washing for removal of fecal material from beef. *Journal of Food Protection*, 58: 984–989. doi: 10.4315/0362-028X-58.9.984

Goulter, R. M., Dykes, G. A. & Small, A. 2008. Decontamination of knives used in the meat industry: Effect of different water temperature and treatment time combinations on the reduction of bacterial numbers on knife surfaces. *Journal of Food Protection*, 71(7): 1338–1342. doi: 10.4315/0362-028X-71.7.1338

Greer, G. G. & Dilts, B. D. 1988. Bacteriology and retail case life of spray-chilled pork. *Canadian Institute of Food Science and Technology Journal*, 21(3): 295–299. doi: 10.1016/S0315-5463(88)70820-2c

Greig, J. D., Waddell, L., Wilhelm, B., Wilkins, W., Bucher, O., Parker, S. & Rajić, A. 2012. The efficacy of interventions applied during primary processing on contamination of beef carcasses with *Escherichia coli*: A systematic review-meta-analysis of the published research. *Food Control*, 27: 385–397. doi: 10.1016/j.foodcont.2012.03.019

Gunes, G., Ozturk, A., Yilmaz, N. & Ozcelik, B. 2011. Maintenance of safety and quality of refrigerated ready-to-cook seasoned ground beef product (meatball) by combining gamma irradiation with modified atmosphere packaging. *Journal of Food Science*, 76: M413-M420. doi: 10.1111/j.1750-3841.2011.02244.x

Gunn, G. J., McKendrick, J., Ternent, H. E., Thomson-Carterd, F., Foster, G. & Synge, B. A. 2007. An investigation of factors associated with the prevalence of verocytotoxin producing *Escherichia coli* O157 shedding in Scottish beef cattle. *The Veterinary Journal*, 174, 554-564. doi: 10.1016/j.tvjl.2007.08.024

Hales, K. E., Wells, J. E., Berry, E. D., Kalchayanand, N., Bono, J. L. & Kim, M. 2017. The effects of monensin in diets fed to finishing beef steers and heifers on growth performance and fecal shedding of *Escherichia coli* O157:H7. *Journal of Animal Science*, 95: 3738–3744. doi: 10.2527/jas.2017.1528

Hancock, D. D., Besser, T. E., Rice, D. H., Ebel, E. D., Herriott, D. E. & Carpenter, L. V. 1998. Multiple sources of *Escherichia coli* O157 in feedlots and dairy farms in the Northwestern USA. *Preventative Veterinary Medicine*, 35: 11–19. doi: 10.1016/s0167-5877(98)00050-6

Haus-Cheymol, R., Espie, E., Che, D., Vaillant, V., De Valk, H. & Desenclos, J. C. 2006. Association between indicators of cattle density and incidence of paediatric haemolytic - uraemic syndrome (HUS) in children under 15 years of age in France between 1996 and 2001: an ecological study. *Epidemiology and Infection*, 134(4): 712–718. doi: 10.1017/S095026880500542X

Heller, C. E., Scanga, J. A., Sofos, J. N., Belk, K. E., Warren-Serna, W., Bellinger, G. R., Bacon, R. T., Rossman, M. L. & Smith, G. C. 2007. Decontamination of beef subprimal cuts intended for blade tenderization or moisture enhancement. *Journal of Food Protection*, 70: 1174–1180. doi: 10.4315/0362-028x-70.5.1174

Herold, S., Karch, H. & Schmidt, H. 2004. Shiga toxin-encoding bacteriophages-genomes in motion. *International Journal of Medical Microbiology*, 294(2-3): 115–121. doi: 10.1016/j.ijmm.2004.06.023

Hochreutener, M., Zweifel, C., Corti, S. & Stephan, R. 2017. Effect of a commercial steam-vacuuming treatment implemented after slaughtering for the decontamination of cattle carcasses. *Italian Journal of Food Safety*, 6: 6864. doi: 10.4081/ijfs.2017.6864

Hong, Y., Pan, Y. & Ebner, P. D. 2014. Meat science and muscle biology symposium: Development of bacteriophage treatments to reduce *Escherichia coli* O157:H7 contamination of beef products and produce. *Journal of Animal Science*, 92: 1366–1377. doi: 10.2527/jas.2013-7272

Horchner, P., Huynh, L., Sumner, J., Vanderlinde, P. B. & Jenson, I. 2020. Performance metrics for slaughter and dressing hygiene at Australian beef export establishments. *Journal of Food Protection*, 83: 996–1001. doi: 10.4315/jfp-19-591

Hovde, C. J., Austin, P. R., Cloud, K. A., Williams, C. J. & Hunt, C. W. 1999. Effect of cattle diet on *Escherichia coli* O157:H7 acid resistance. *Applied and Environmental Microbiology*, 65(7): 3233–3235. doi: 10.1128/AEM.65.7.3233-3235.1999

Hsu, H., Sheen, S., Sites, J., Cassidy, J., Scullen, B. & Sommers, C. 2015. Effect of high pressure processing on the survival of Shiga toxin-producing *Escherichia coli* (Big Six vs. O157:H7) in ground beef. *Food Microbiology* 48: 1–7. doi: 10.1016/j.fm.2014.12.002

Hudson, J. A., Billington, C., Cornelius, A. J., Wilson, T., On, S. L. W., Premaratine, A. & King, N. J. 2013. Use of a bacteriophage to inactivate *Escherichia coli* O157:H7 on beef. *Food Microbiology*, 36: 14–21. doi: 10.1016/j.fm.2013.03.006.

Huffman, R. D. 2002. Current and future technologies for the decontamination of carcasses and fresh meat. *Meat Science*, 62(3): 285–294. doi: 10.1016/S0309-1740(02)00120-1

International Commission on Microbiological Specifications for Foods (ICMSF). 2011. Microorganisms in Foods 8: Use of data for assessing process control and product acceptance. New York, NY: Springer.

Jacob, M. E., Fox, J. T., Narayanan, S. K., Drouillard, J. S., Renter, D. G. & Nagaraja, T. G. 2008. Effects of feeding wet corn distiller's grains with solubles with or without monensin and tylosin on the prevalence and antimicrobial susceptibilities of fecal food-borne pathogenic and commensal bacteria in feedlot cattle. *Journal of Animal Science*, 86: 1182–1190. doi: 10.2527/jas.2007-0091

Jacob, M. E., Callaway, T. R. & Nagaraja, T. G. 2009. Dietary interactions and interventions affecting *Escherichia coli* O157 colonization and shedding in cattle. *Foodborne Pathogens and Disease*, 6: 785–792. doi: 10.1089/fpd.2009.0306

Jacob, M. E., Almes, K. M., Shi, X., Sargeant, J. M. & Nagaraja, T. G. 2011. *Escherichia coli* O157:H7 genetic diversity in bovine fecal samples. *Jounral of Food Protection*, 74(7): 1186–1188. doi: 10.4315/0362-028X.JFP-11-022

Jericho, K. W., Bradley, J. A. & Kozub, G. C. 1995. Microbiologic evaluation of carcasses before and after washing in a beef slaughter plant. *Journal of the American Veterinary Medical Association*, 206(4): 452–5. PMID: 7632246. https://pubmed.ncbi.nlm.nih.gov/7632246

Jericho, K. W. F., Kozub, G. C., Bradley, J. A., Gannon, V. P. J., Golsteyn-Thomas, E. J., Gierus, M., Nishiyama, B. J., King, R. K., Tanaka, E. E., D'Souza, S. & Dixon-MacDougall, J. M. 1996. Microbiological verification of the control of the processes of dressing, cooling and processing of beef carcasses at a high line-speed abattoir. *Food Microbiology*, 13(4): 291–301. doi: 10.1006/fmic.1996.0035

Jiang, Y., Scheinberg, J. A., Senevirathne, R. & Cutter, C. N. 2015. The efficacy of short and repeated high-pressure processing treatments on the reduction of non-O157:H7 Shiga-toxin producing *Escherichia coli* in ground beef patties. *Meat Science*, 102: 22–26. doi: 10.1016/j.meatsci.2014.12.001

Johnson, J. Y. M., Thomas, J. E., Graham, T. A., Townshend, I., Byrne, J., Selinger, L. B. & Gannon, V. P. 2003. Prevalence of *Escherichia coli* O157:H7 and *Salmonella* spp. in surface waters of southern Alberta and its relation to manure sources. *Canadian Journal of Microbiology*, 49(5): 326–335. doi: 10.1139/w03-046

Jordan, D. & McEwen, S. A. 1998. Effect of duration of fasting and a short-term high-roughage ration on the concentration of *Escherichia coli* biotype 1 in cattle feces. *Journal of Food Protection*, 61(5): 531-534. doi: 10.4315/0362-028x-61.5.531

Kalchayanand, N., Arthur, T. M., Bosilevac, J. M., Brichta-Harhay, D. M., Guerini, M. N., Wheeler, T. L. & Koohmaraie, M. 2008. Evaluation of various antimicrobial interventions for the reduction of *Escherichia coli* O157:H7 on bovine heads during processing. *Journal of Food Protection*, 71(3): 621–624. doi: 10.4315/0362-028X-71.3.621

Kalchayanand, N., Arthur, T. M., Bosilevac, J. M., Schmidt, J. W., Wang, R., Shackelford, S. D. & Wheeler, T. L. 2012. Evaluation of commonly used antimicrobial interventions for fresh beef inoculated with Shiga toxin-producing *Escherichia coli* serotypes O26, O45, O103, O111, O121, O145, and O157:H7. *Journal of Food Protection*, 75: 1207–1212. doi: 10.4315/0362-028x.jfp-11-53

Kalchayanand, N., Arthur, T. M., Bosilevac, J. M., Schmidt, J. W., Wang, R., Shackelford, S. & Wheeler, T. L. 2015. Efficacy of antimicrobial compounds on surface decontamination of seven Shiga toxin–producing *Escherichia coli* and *Salmonella* inoculated onto fresh beef. *Journal of Food Protection*, 78: 503–510. doi: 10.4315/0362-028x.jfp-14-268

Kang, S., Ravensdale, J., Coorey, R., Dykes, G.A. & Barlow, R. 2019. A comparison of 16S rRNA profiles through slaughter in Australian export beef abattoirs. *Frontiers in Microbiology*, 10: 2747. doi: 10.3389/fmicb.2019.02747

Keen, J. E. & Elder, R. O. 2002. Isolation of Shiga-toxigenic *Escherichia coli* O157 from hide surfaces and the oral cavity of finished beef feedlot cattle. *Journal of the American Veterinary Medical Association*, 220(6): 756–763. doi: 10.2460/javma.2002.220.756

Kennedy, T. G., Giotis, E. S. & McKevitt, A. I. 2014. Microbial assessment of an upward and downward dehiding technique in a commercial beef processing plant. *Meat Science*, 97: 486–489. doi: 10.1016/j.meatsci.2014.03.009

Kenney, P. B., Prasai, R. K., Cazpbell, R. E. Kastner, C. L. & Fung, D. Y. C. 1995. Microbiological quality of beef carcasses and vacuum-packaged subprimals: process intervention during slaughter and fabrication. *Journal of Food Protection*, 58(6): 633–638. doi: 10.4315/0362-028X-58.6.633

Kerner, K., Bridger, P. S., Köpf, G., Fröhlich, J., Barth, S., Willems, H., Bauerfeind, R., Baljer, G. & Menge, C. 2015. Evaluation of biological safety in vitro and immunogenicity in vivo of recombinant *Escherichia coli* Shiga toxoids as candidate vaccines in cattle. *Veterinary Research*, 46(1): 38. doi: 10.1186/s13567-015-0175-2

Kimmitt, P. T., Harwood, C. R. & Barer, M. R. 2000. Toxin gene expression by Shiga toxin-producing *Escherichia coli*: the role of antibiotics and the bacterial SOS response. *Emerging Infectious Diseases*, 6: 458–465. doi: 10.3201/eid0605.000503

King, T., Kocharunchitt, C., Gobius, K., Bowman, J. P. & Ross, T. 2016. Physiological response of *Escherichia coli* O157:H7 Sakai to dynamic changes in temperature and water activity as experienced during carcass chilling. *Molecular and Cellular Proteomics*, 15(11): 3331–3347. doi: 10.1074/mcp.M116.063065

Kinsella, K. J., Sheridan, J. J., Rowe, T. A., Butler, F., Delgado, A., Quispe-Ramirez, A., Blair, I. S. & McDowell, D. A. 2006. Impact of a novel spray-chilling system on surface microflora, water activity and weight loss during beef carcass chilling. *Food Microbiology*, 23(5): 483–490. doi: 10.1016/j.fm.2005.05.013

Kirsch, K. R., Tolen, T. N., Hudson, J. C., Castillo, A., Griffin, D. & Taylor, T. M. 2017. Effectiveness of a commercial lactic acid bacteria intervention applied to inhibit Shiga toxin-producing *Escherichia coli* on refrigerated vacuum-aged beef. *International Journal of Food Science*, 8070515. doi: 10.1155/2017/8070515

Kocharunchitt, C., Mellefont, L., Bowman, J. P. & Ross, T. 2020. Application of chlorine dioxide and peroxyacetic acid during spray chilling as a potential antimicrobial intervention for beef carcasses. *Food Microbiology*, 87: 103355. doi: 10.1016/j.fm.2019.103355

Koh, E. L. 2020. The impact of commercial freezing on Top 7 STEC detections in New Zealand bovine bulk manufacturing meat cartons. *Department of Faculty of Veterinary Science, University of Sydney*. (Master Thesis).

Köhler, B., Karch, H. & Schmidt, H. 2000. Antibacterials that are used as growth promoters in animal husbandry can affect the release of Shiga-toxin-2-converting bacteriophages and Shiga toxin 2 from *Escherichia coli* strains. *Microbiology*, 146: 1085–1090. doi: 10.1099/00221287-146-5-1085

Kudra, L. L., Sebranek, J.G., Dickson, J. S., Mendonca, A. F., Larson, E. M., Jackson-Davis A. L. & Lu, Z. 2011. Effects of vacuum or modified atmosphere packaging in combination with irradiation for control of *Escherichia coli* O157:H7 in ground beef patties. *Journal of Food Protection*, 74: 2018–2023. doi: 10.4315/0362-028X. JFP-11-289

Kundu, D., Gill, A., Lui, C., Goswami, N. & Holley, R. 2014. Use of low dose e-beam irradiation to reduce *E. coli* O157:H7, non-O157 (VTEC) *E. coli* and *Salmonella* viability on meat surfaces. *Meat Science*, 96: 413–418. doi: 10.1016/j. meatsci.2013.07.034

Lan, H., Hosomi, K. & Kunisawa, J. 2019. *Clostridium perfringens* enterotoxin-based protein engineering for the vaccine design and delivery system. *Vaccine*, 37(42): 6232–6239. doi: 10.1016/j.vaccine.2019.08.032

LeJeune, J. T., Besser, T. E. & Hancock, D. D. 2001. Cattle water troughs as reservoirs of *Escherichia coli* O157. *Applied Environmental Microbiology*, 67: 3053–57. doi: 10.1128/AEM.67.7.3053-3057.2001

LeJeune, J. T., Besser, T. E., Rice, D. H., Berg, J. L., Stilborn, R. P. & Hancock, D. D. 2004. Longitudinal study of fecal shedding of *Escherichia coli* O157:H7 in feedlot cattle: predominance and persistence of specific clonal types despite massive cattle population turnover. *Applied and Environmental Microbiology*, 70: 377–84. doi: 10.1128/aem.70.1.377-384

Li, S. L., Kundu, D. & Holley, R. A. 2015. Use of lactic acid with electron beam irradiation for control of *Escherichia coli* O157:H7, non-O157 VTEC *E. coli*, and *Salmonella* serovars on fresh and frozen beef. *Food Microbiology*, 46: 34–39. doi: 10.1016/j. fm.2014.06.018

Liao, Y. T., Brooks, J. C., Martin, J. N., Echeverry, A., Loneragan, G. H. & Brashears, M. M. 2015. Antimicrobial interventions for O157:H7 and Non-O157 Shiga toxin-producing *Escherichia coli* on beef subprimal and mechanically tenderized steaks. *Journal of Food Protection*, 78(3): 511–517. doi: 10.4315/0362-028X.JFP-14-178

Liu, H., Niu, Y. D., Meng, R., Wang, J., Li, J., Johnson, R. P., McAllister, T. A. & Stanford, K. 2015. Control of *Escherichia coli* O157 on beef at 37, 22 and 4°C by T5-, T1-, T4-and O1-like bacteriophages. *Food Microbiology*, 51: 69–73. doi: 10.1016/j.fm.2015.05.001

Logue, C. M., Sheridan, J. J. & Harrington, D. 2005. Studies of steam decontamination of beef inoculated with *Escherichia coli* O157:H7 and its effect on subsequent storage. *Journal of Applied Microbiology*, 98: 741–751. doi: 10.1111/j.1365-2672.2004. 02511.x

Mather, A. E., Innocent, G. T., McEwen, S. A., Reilly, W. J., Taylor, D. J., Steele, W. B., Gunn, G. J., Ternent, H. E., Reid, S. W. J. & Mellor, D. J. 2007. Risk factors for hide contamination of Scottish cattle at slaughter with *Escherichia coli* O157. *Preventive Veterinary Medicine*, 80:257–270. doi: 10.1128/AEM.00770-08

Mather, A. E., Reid, S. W. J., McEwen, S. A., Ternent, H. E., Reid-Smith, R. J., Boerlin, P., Taylor, D. J., Steele, W. B., Gunn, G. J. & Mellor, D. J. 2008. Factors associated with cross-contamination of hides of Scottish cattle by *Escherichia coli* O157. *Applied and Environmental Microbiology*, 74(20): 6313–6319. doi: 10.1128/AEM.00770-08

Martorelli, L., Garbaccio, S., Vilte, D. A., Albanese, A. A., Mejías, M. P., Palermo, M. S., Mercado, E. C., Ibarra, C. E. & Cataldi, A. A. 2017. Immune response in calves vaccinated with type three secretion system antigens and Shiga toxin 2B subunit of *Escherichia coli* O157:H7. *PLOS ONE*, 12(1): e0169422. doi: 10.1371/journal.pone.0169422

McAllister, T. A., Bach, S. J., Stanford, K. & Callaway, T. R. 2006. Shedding of *Escherichia coli* O157:H7 by cattle fed diets containing monensin or tylosin. *Journal of Food Protection*, 69: 2075–2083. doi: 10.4315/0362-028x-69.9.2075

McCann, M. S., McGovern, A. C., McDowell, D. A., Blair, L. S. & Sheridan, J. J. 2006. Surface decontamination of beef inoculated with *Salmonella* Typhimurium DT104 or *Escherichia coli* O157:H7 using dry air in a novel heat treatment apparatus. *Journal of Applied Microbiology*, 101: 1177–1187. doi: 10.1111/j.1365-2672.2006.02988.x

McEvoy, J. M., Doherty, A. M., Finnerty, M., Sheridan, J. J., McGuire, L., Blair, I. S., McDowell, D. A. & Harrington, D. 2001. The relationship between hide cleanliness and bacterial numbers on beef carcasses at a commercial abattoir. *Letters in Applied Microbiology*, 30: 390–395. doi: 10.1046/j.1472-765x.2000.00739.x

McEvoy, J. M., Doherty, A. M., Sheridan, J. J. & McGuire, L. 1999. Baseline study of the microflora of beef carcasses in a commercial abattoir. *Irish Journal of Agricultural and Food Research (Abstract)*, 38(1): 157.

McEvoy, J. M., Sheridan, J. J., Blair, I. S. & McDowell, D. A. 2004. Microbial contamination on beef in relation to hygiene assessment based on criteria used in EU Decision 2001/471/EC. *International Journal of Food Microbiology*, 92(2): 217–225. doi: 10.1016/j.ijfoodmicro.2003.09.010

McMillin, K. & Michel, M. 2000. Reduction of *E. coli* in ground beef with gaseous ozone. *Louisiana Agriculture*, 43: 35. www.researchgate.net/publication/285663498_Reduction_of_E_coli_in_ground_beef_with_gaseous_ozone

MLA (Meat & Livestock Australia). No date. Organic Acids. Cited on 18 August 2022. www.mla.com.au/globalassets/mla-corporate/research-and-development/program-areas/food-safety/documents/food-safety-intervention/organic-acids.pdf

Mellefont, L. A., Kocharunchitt, C. & Ross, T. 2015. Combined effect of chilling and desiccation on survival of *Escherichia coli* suggests a transient loss of culturability. *International Journal of Food Microbiology*, 208: 1–10. doi: 10.1016/j.ijfoodmicro.2015.04.024

Mellor, G. E., Fegan, N., Duffy, L. L., McMillan, K. E., Jordan, D. & Barlow, R. S. 2016. National survey of Shiga toxin–producing *Escherichia coli* serotypes O26, O45,

O103, O111, O121, O145, and O157 in Australian beef cattle feces. *Journal of Food Protection*, 79(11): 1868–1874. doi: 10.4315/0362-028X.JFP-15-507

Mies, P. D., Covington, B. R., Harris, K. B., Lucia, L. M., Acuff, G. R. & Savell, J. W. 2004. Decontamination of cattle hides prior to slaughter using washes with and without antimicrobial agents. *Journal of Food Protection,* 67: 579–582. doi: 10.4315/0362-028x-67.3.579

Milios, K., Drosinos, E. H. & Zoiopoulos, P. E. 2017. Carcass decontamination methods in slaughterhouses: A review. *Journal of the Hellenic Veterinary Medical Society*, 65(2): 65–78. doi: 10.12681/jhvms.15517

Minihan, D., O'Mahony, M., Whyte, P. & Collins, J. D. 2003. An investigation on the effect of transport and lairage on the faecal shedding prevalence of *Escherichia coli* O157 in cattle. *Journal of Veterinary Medicine,* 50: 378–382. doi: 10.1046/j.1439-0450.2003.00674.x

Miszczycha, S. D., Perrin, F., Ganet, S., Jamet, E., Tenehaus-Aziza, F., Montel, M. & Theenot-Sergentet, D. 2013. Behavior of different Shiga toxin-producing *Escherichia coli* serotypes in various experimentally contaminated raw-milk cheeses. *Applied Environmental Microbiology,* 79: 150–158. doi: 10.1128/AEM.02192-12

Miszczycha, S. D., Bel, N., Gay-Perret, P., Michel, V., Montel, M. C. & Sergentet-Thevenot, D. 2016. Short communication: Behavior of different Shiga toxin-producing *Escherichia coli* serotypes (O26:H11, O103:H2, O145:H28, O157:H7) during the manufacture, ripening, and storage of a white mold cheese. *Journal of Dairy Science*, 99(7): 5224–5229. doi: 10.3168/jds.2015-10803

Money, P., Kelly, A. F., Gould, S. W., Denholm-Price, J., Threlfall, E. J. & Fielder, M. D. 2010. Cattle, weather and water: mapping *Escherichia coli* O157:H7 infections in humans in England and Scotland. *Environmental Microbiology*, 12: 2633–44. doi: 10.1111/j.1462-2920.2010.02293.x

Morales, P., Calzada, J., Ávila, M. & Nuñez, M. 2008. Inactivation of *Escherichia coli* O157:H7 in ground beef by single-cycle and multiple-cycle high-pressure treatments. *Journal of Food Protection,* 71: 811–815. doi: 10.4315/0362-028x-71.4.811

Moxley, R. A. & Acuff, G. R. 2014. Peri- and postharvest factors in the control of Shiga toxin-producing *Escherichia coli* in beef. *Microbiology Spectrum,* 2(6). doi: 10.1128/microbiolspec.EHEC-0017-2013

Munns, K. D., Selinger, L., Stanford, K., Selinger, L. B. & McAllister, T. A. 2014. Are super-shedder feedlot cattle really super? *Foodborne Pathogen and Disease,* 11: 329–331. doi: 10.1089/fpd.2013.1621

Munns, K. D., Zaheer, R., Xu, Y., Stanford, K., Laing, C. R., Gannon, V. P. J., Selinger, L. B. & McAllister, T. A. 2015. Comparative genomics of *Escherichia coli* O157:H7 isolated from super-shedder and low-shedder cattle. *PLOS ONE*, 11: e0151673. doi: 10.1371/ journal.pone.0151673

Muriana, P., Eager, J., Wellings, B., Morgan, B., Nelson, J. & Kushwaha, K. 2019. Evaluation of antimicrobial interventions against *E. coli* O157:H7 on the surface of raw beef to reduce bacterial translocation during blade tenderization. *Foods,* 8: 80. doi: 10.3390/foods8020080

Nastasijevic, I., Mitrovic, R. & Buncic, S. 2008. Occurrence of *Escherichia coli* O157 on hides of slaughtered cattle. *Letters in Applied Microbiology,* 46: 126–131. doi: 10.1111/j.1472-765X.2007.02270.x

Nortjé, G. L. & Naudé, R. T. 1981. Microbiology of beef carcass surfaces. *Journal of Food Protection,* 44(5): 355–358. doi: 10.4315/0362-028X-44.5.355

Nou, X., Rivera-Betancourt, M., Bosilevac, J. M., Wheeler, T. L., Shackelford, S. D., Gwartney, B. L., Reagan, J. O. & Koohmaraie, M. 2003. Effect of chemical dehairing on the prevalence of *Escherichia coli* O157:H7 and the lvels of aerobic bacteria and *Enterobacteriaceae* on carcasses in a commercial beef processing plant. *Journal of Food Protection,* 66(11): 2005–2009. doi: 10.4315/0362-028X-66.11.2005

Nutsch, A. L., Phebus, R. K., Riemann, M. J., Schafer, D. E., Boyer, J. E., Wilson, R. C., Leising, J. D. & Kastner, C. L. 1997. Evaluation of a steam pasteurization process in a commercial beef processing facility. *Journal of Food Protection,* 60(5): 485–492. doi: 10.4315/0362-028X-60.5.485

Ogden, I. D., Hepburn, N. F., MacRae, M., Strachan, N. J. C., Fenlon, D. R., Rusbridge, S. M. & Pennington, T. H. 2002. Long-term survival of *Escherichia coli* O157 on pasture following an outbreak associated with sheep at a scout camp. *Letters in Applied Microbiology,* 34(2): 100–104. doi: 10.1046/j.1472-765x.2002.01052.x

Paddock, Z. D., Walker, C. E., Drouillard, J. S. & Nagaraja, T. G. 2011. Dietary monensin level, supplemental urea, and ractopamine on fecal shedding of *Escherichia coli* O157:H7 in feedlot cattle. *Journal of Animal Science,* 89: 2829–2835. doi: 10.2527/jas.2010-3793

Paddock, Z. D., Renter, D. G., Shi, X., Krehbiel, C. R., DeBey, B. & Nagaraja, T. G. 2013. Effects of feeding dried distillers grains with supplemental starch on fecal shedding of *Escherichia coli* O157:H7 in experimentally inoculated steers. *Journal of Animal Science,* 91(3): 1362–1370. doi: 10.2527/jas.2012-5618

Patterson, M. F. & Kilpatrick, D. J. 1998. The combined effect of high hydrostatic pressure and mild heat on inactivation of pathogens in milk and poultry. *Journal of Food Protection,* 61: 432–436. doi: 10.4315/0362-028X-61.4.432

Peng, S., Hoffmann, W., Bockelmann, W., Hummerjohann, J., Stephan, R. & Hammer, P. 2013a. Fate of Shiga toxin-producing and generic *Escherichia coli* during production and ripening of semihard raw milk cheese. *Journal of Dairy Science,* 96: 815–23. doi: 10.3168/jds.2012-5865

Peng, S., Schafroth, K., Jakob, E., Stephan, R. & Hummerjohann, J. 2013b. Behaviour of *Escherichia coli* strains during semi-hard and hard raw milk cheese production. *International Dairy Journal,* 31: 117–120. doi: 10.1016/j.idairyj.2013.02.012

Perrin, F., Tenenhaus-Aziza, F., Michel, V., Miszczycha, S., Bel, N. & Sanaa, M. 2015. Quantitative risk assessment of haemolytic and uremic syndrome linked to O157:H7 and non-O157:H7 Shiga-toxin-producing *Escherichia coli* strains in raw milk soft cheeses. *Risk Analysis,* 35: 109–128. doi: 10.1111/risa.12267

Phebus, R. K., Nutsch, A. L., Schafer, D. E., Wilson, R. C., Riemann, M. J., Leising, J. D., Kastner, C. L., Wolf, J. R. & Prasai, R. K. 1997. Comparison of steam pasteurization and other methods for reduction of pathogens on surfaces of freshly slaughtered beef. *Journal of Food Protection,* 60: 476–484. doi: 10.4315/0362-028x-60.5.476

Phetxumphou, K. 2018. Novel Approaches to Exposure Assessment and Dose Response to Contaminants in Drinking Water and Food. https://vtechworks.lib.vt.edu/bitstream/handle/10919/94582/Phetxumphou_K_D_2018.pdf?isAllowed=y&sequence=1

Pointon, A., Kiermeier, A. & Fegan, N. 2012. Review of the impact of pre-slaughter feed curfews of cattle, sheep and goats on food safety and carcase hygiene in Australia. *Food Control,* 26: 313–321. doi: 10.1016/j.foodcont.2012.01.034

Prasai, R. K., Phebus, R. K., Garcia Zepeda, C. M., Kastner, C. L., Boyle, A. E. & Fung, D. Y. C. 1995. Effectiveness of trimming and/or washing on microbiological quality of beef carcasses. *Journal of Food Protection,* 58: 1114–1117. doi: 10.4315/0362-028x-58.10.1114

Ransom, J., Belk, K., Sofos, J., Stopforth, J., Scanga, J. & Smith, G. 2003. Comparison of intervention technologies for reducing *Escherichia coli* O157:H7 on beef cuts and trimmings. *Food Protection Trends,* 23: 24–34.

Sabouri, S., Sepehrizadeh, Z., Amirpour-Rostami, S. & Skurnik, M. 2017. A minireview on the in vitro and in vivo experiments with anti-*Escherichia coli* O157:H7 phages as potential biocontrol and phage therapy agents. *International Journal of Food Microbiology,* 243: 52–57. doi: 10.1016/j.ijfoodmicro.2016.12.004

Salim, A. P. A. A., Canto, A. C. V. C. S., Costa-Lima, B. R. C., Simoes, J. S., Panzenhagen, P. H. N., Franco, R. M., Silva, T. J. P. & Conte-Junior, C. A. 2017. Effect of lactic acid on *Escherichia coli* O157:H7 and on color stability of vacuum-packaged beef steaks under high storage temperature. *Journal of Microbiology, Biotechnology and Food Sciences,* 6(4): 1054–1058. doi: 10.15414/jmbfs.2017.6.4.1054-1058

Sanderson, M. W., Sargeant, J. M., Shi, X., Nagaraja, T. G., Zurek, L. & Alam, M. J. 2006. Longitudinal emergence and distribution of *Escherichia coli* O157 genotypes in a beef feedlot. *Applied and Environmental Microbiology,* 72: 7614–7619. doi: 10.1128/AEM.01412-06

Schamberger, G. P., Phillips, R. L., Jacobs, J. L. & Diez-Gonzalez, F. 2004. Reduction of *Escherichia coli* O157:H7 populations in cattle by addition of colicin E7-producing *E. coli* to feed. *Applied and Environmental Microbiology,* 70(10): 6053–6060. doi: 10.1128/AEM.70.10.6053-6060.2004

Schmidt, J. W., Wang, R., Kalchayanand, N., Wheeler, T. L. & Koohmaraie, M. 2012. Efficacy of hypobromous acid as a hide-on carcass antimicrobial intervention. *Journal of Food Protection*, 75(5): 955–958. doi: 10.4315/0362-028X.JFP-11-433

Schneider, L. G., Stromberg, Z. R., Lewis, G. L., Moxley, R. A. & Smith, R. 2018. Cross-sectional study to estimate the prevalence of enterohaemorrhagic *Escherichia coli* on hides of market beef cows at harvest. *Zoonoses Public Health*, 65: 625–636. doi: 10.1111/zph.12468

Schnell, T. D., Sofos, J. N., Littlefield, V. G., Morgan, J. B., Gorman, B. M., Clayton, R. P. & Smith, G. C. 1995. Effects of postexsanguination dehairing on the microbial load and visual cleanliness of beef carcasses. *Journal of Food Protection*, 58(12): 1297–1302. doi: 10.4315/0362-028X-58.12.1297

Schuehle-Pfeiffer, C. E., King, D. A., Lucia, L. M., Cabrera Diaz, E., Acuff, G. R., Randel, R. D., Welsh, T.H., Oliphint, C. & Vann, S. 2009. Influence of transportation stress and animal temperament on fecal shedding of *Escherichia coli* O157:H7 in feedlot cattle. *Meat Science*, 81: 300–306. doi: 10.1016/j.meatsci.2008.08.005

Schulz, S., Stephan, A., Hahn, S., Bortesi, L., Jarczowski, F., Bettmann, U., Paschke, A. K., Tuse, D., Stahl, C. H., Giritch, A. & Gleba, Y. 2015. Broad and efficient control of major foodborne pathogenic strains of *Escherichia coli* by mixtures of plant-produced colicins. *Proceedings of the National Academy of Sciences (USA)*, 112: E5454–60. doi: 10.1073/pnas.1513311112

Sheridan, J. 1998. Sources of contamination during slaughter and measures for control. *Journal of Food Safety*, 18(4): 321–339. doi: 10.1111/j.1745-4565.1998.tb00223.x

Small, A., Reid, C. A. & Buncic, S. 2003. Conditions in lairages at abattoirs for ruminants in southwest England and in vitro survival of *Escherichia coli* O157, *Salmonella* Kedougou, and *Campylobacter jejuni* on lairage-related substrates. *Journal of Food Protection*, 66: 1570–1575. doi: 10.4315/0362-028x-66.9.1570

Small, A., Wells-Burr, B. & Buncic, S. 2005. An evaluation of selected methods for the decontamination of cattle hides prior to skinning. *Meat Science*, 69(2): 263–268. doi: 10.1016/j.meatsci.2004.07.005

Small, A., James, C., Purnell, G., Losito, P., James, S. & Buncic, S. 2007. An evaluation of simple cleaning methods that may be used in red meat abattoir lairages. *Meat Science*, 75: 220–228. doi: 10.1016/j.meatsci.2006.07.007

Smith, M. G. & Graham, A. 1978. Destruction of *Escherichia coli* and *Salmonellae* on mutton carcases by treatment with hot water. *Meat Science*, 2(2): 119–128. doi: 10.1016/0309-1740(78)90012-8

Smith, M. G. 1992. Destruction of bacteria on fresh meat by hot water. *Epidemiology and Infection*, 109: 491-496. doi: 10.1017/s0950268800050482

Smith, D., Blackford, M., Younts, S., Moxley, R., Gray, J., Hungerford, L., Milton, T. & Klopfenstein, T. 2001. Ecological relationships between the prevalence of cattle shedding

Escherichia coli O157:H7 and characteristics of the cattle or conditions of the feedlot pen. *Journal of Food Protection,* 64: 1899–903. doi: 10.4315/0362-028x-64.12.1899

Smith, D. R., Klopfenstein, T., Moxley, R., Milton, C. T., Hungerford, L. & Gray, J. T. 2002. An evaluation of three methods to clean feedlot water tanks. *The Bovine Practitioner,* 36: 1–4. doi: 10.21423/bovine-vol36no1p1-4

Smith, L., Mann, J. E., Harris, K., Miller, M. F. & Brashears, M. M. 2005a. Reduction of *Escherichia coli* O157:H7 and *Salmonella* in ground beef using lactic acid bacteria and the impact on sensory properties. *Journal of Food Protection,* 68: 1587–1592. doi: 10.4315/0362-028x-68.8.1587

Smith, D. R., Moxley, R. A., Clowser, S. L., Folmer, J. D., Hinkley, S., Erickson, G. E. & Klopfenstein, T. J. 2005b. Use of rope devices to describe and explain the feedlot ecology of *Escherichia coli* O157:H7 by time and place. *Foodborne Pathogens and Disease,* 2: 50–60. doi: 10.1089/fpd.2005.2.50

Snedeker, K. G., Campbell, M. & Sargeant, J. M., 2012. A systematic review of vaccinations to reduce the shedding of Escherichia coli O157 in the faeces of domestic ruminants. *Zoonoses and Public Health,* 59: 126–138. doi: 10.1111/j.1863-2378.2011.01426.x

Solomakos, N., Govaris, A., Koidis, P. & Botsoglou, N. 2008. The antimicrobial effect of thyme essential oil, nisin and their combination against *Escherichia coli* O157:H7 in minced beef during refrigerated storage. *Meat Science,* 80: 159–166. doi: 10.1016/j.meatsci.2007.11.014

Solomakos, N., Govari, M., Botsoglou, E. & Pexara, A. 2019. Effect of modified atmosphere packaging on physicochemical and microbiological characteristics of *Graviera Agraphon* cheese during refrigerated storage. *Journal of Dairy Research,* 86: 483–489. doi: 10.1017/S0022029919000724

Sommers, C., Rajkowski, K. T., Scullen, O. J., Cassidy, J., Fratamico, P. & Sheen, S. S. 2015. Inactivation of Shiga toxin-producing *Escherichia coli* in lean ground beef by gamma irradiation. *Food Microbiology,* 49: 231–234. doi: 10.1016/j.fm.2015.02.013

Speranza, B., Liso, A., Russo, V. & Corbo, M. R. 2020. Evaluation of the potential of biofilm formation of *Bifidobacterium longum* subsp. *infantis* and *Lactobacillus reuteri* as competitive biocontrol agents against pathogenic and food spoilage bacteria. *Microorganisms,* 8: 177–191. doi: 10.3390/microorganisms8020177

Stacey, K. F., Parsons, D. J., Christiansen, K. H. & Burton, C. H. 2007. Assessing the effect of interventions on the risk of cattle and sheep carrying *Escherichia coli* O157:H7 to the abattoir using a stochastic model. *Preventative Veterinary Medicine,* 79: 32–45. doi: 10.1016/j.prevetmed.2006.11.007

Stanford, K., Bryan, M., Peters, J., Gonzalez, L. A., Stephens, T. P. & Schwartzkopf-Genswein, K. S. 2011. Effects of long- or short-haul transportation of slaughter heifers and cattle liner microclimate on hide contamination with *Escherichia coli* O157. *Journal of Food Protection,* 74: 1605–1610. doi: 10.4315/0362-028X.JFP-11-154

Stenkamp-Strahm, C., Lombard, J. E., Magnuson R. J., Linke, L. M., Magzamen, S., Urie, N. J., Shivley, C. B. & McConnel, C. S. 2018. Preweaned heifer management on US dairy operations: Part IV. Factors associated with the presence of *Escherichia coli* O157 in preweaned dairy heifers. *Journal of Dairy Science*, 101: 9214–9228. doi: 10.3168/jds.2018-14659

Stephens, T. P., Loneragan, G. H., Karunasena, E. & Brashears, M. M. 2007. Reduction of *Escherichia coli* O157 and *Salmonella* in feces and on hides of feedlot cattle using various doses of a direct-fed microbial. *Journal of Food Protection*, 70(10): 2386–2391. doi: 10.4315/0362-028X-70.10.2386

Stevenson, S. M. L., Cook, S. R., Bach, S. J. & McAllister, T. A. 2004. Effects of storage, water source, bacterial and fecal loads on the efficacy of electrolyzed oxidizing (EO) water for the control of *Escherichia coli* O157:H7. *Journal of Food Protection*, 67: 1377–1383. doi: 10.4315/0362-028x-67.7.1377

Stivarius, M. R., Pohlman, F. W., McElyea, K. S. & Waldroup, A. L. 2002. Effects of hot water and lactic acid treatment of beef trimmings prior to grinding on microbial, instrumental color and sensory properties of ground beef during display. *Meat Science*, 60: 327–334. doi: 10.1016/S0309-1740(01)00127-9

Stopforth, J. D., Lopes, M., Shultz, J. E., Miksch, R. R. & Samadpour, M. 2006. Location of bung bagging during beef slaughter influences the potential for spreading pathogen contamination on beef carcasses. *Journal of Food Protection*, 69: 1452–1455. doi: 10.4315/0362-028x-69.6.1452

Strachan, N. J. C., Dunn, G. M., Locking, M. E., Reid, T. M. S. & Ogden, I. D. 2006. *Escherichia coli* O157: Burger bug or environmental pathogen? *International Journal of Food Microbiology*, 112(2): 129–137. doi: 10.1016/j.ijfoodmicro.2006.06.021

Talley, J. L., Wayadande, A. C., Wasala, L. P., Gerry, A. C., Fletcher, J., DeSilva, U. & Gilliland, S. E. 2009. Association of *Escherichia coli* O157:H7 with filth flies (Muscidae and Calliphoridae) captured in leafy greens fields and experimental transmission of *E. coli* O157:H7 to spinach leaves by house flies (diptera: Muscidae). *Journal of Food Protection*, 72: 1547–1552. doi: 10.4315/0362-028x-72.7.1547

Tanaro, J. D., Piaggio, M. C., Galli, L., Gasparovic, A. M. C., Procura, F., Molina, D. A., Vitón, M., Zolezzi, G. & Rivas, M. 2014. Prevalence of *Escherichia coli* O157:H7 in surface water near cattle feedlots. *Foodborne Pathogens and Disease*, 11(12): 960–965. doi: 10.1089/fpd.2014.1770

Thomas, J. D., Allen, D. M., Hunt, M. C. & Kastner, C. L. 1997. Nutritional regime, post-slaughter conditioning temperature, and vacuum packing effects on bacteriology of beef carcasses and retail meat cuts. *Journal of Food Protection*, 40: 678-682. doi: 10.4315/0362-028X-40.10.678

Tittor, A. W., Tittor, M. G., Brashears, M. M., Brooks, J. C., Garmyn, A. J. & Miller, M. F. 2011. Effects of simulated dry and wet chilling and aging of beef fat and lean tissues on the reduction of *Escherichia coli* O157:H7 and *Salmonella*. *Journal of Food Protection*, 74: 289–293. doi: 10.4315/0362-028X.JFP-10-295

Tolen, T., Xie, Y., Hairgrove, T., Gill, J. & Taylor, T. 2018. Evaluation of commercial prototype bacteriophage intervention designed for reducing O157 and Non-O157 Shiga-toxigenic *Escherichia coli* (STEC) on beef cattle hide. *Foods*, 7(7): 114. doi: 10.3390/foods7070114

Tomat, D., Migliore, L., Aquili, V., Quiberoni, A. & Balagué, C. 2013a. Phage biocontrol of enteropathogenic and Shiga toxin-producing *Escherichia coli* in meat products. *Frontier in Cellular Infection Microbiology*, 3: 20. doi: 10.3389/fcimb.2013.00020

Tomat, D., Mercanti, D., Balagué, C. & Quiberoni, A. 2013b. Phage biocontrol of enteropathogenic and Shiga toxin-producing *Escherichia coli* during milk fermentation. *Letters in Applied Microbiology*, 57: 3–10. doi: 10.1111/lam.12074

Tomat, D., Casabonne, C., Aquilia, V., Balagué, C. & Quiberoni, A. 2018. Evaluation of a novel cocktail of six lytic bacteriophages against Shiga toxin-producing *Escherichia coli* in broth, milk and meat. *Food Microbiology*, 76: 438–442. doi: 10.1016/j.fm.2018.07.006

Tymensen, L., Booker, C. W., Hannon, S. J., Cook, S. R., Zaheer, R., Read, R. & McAllister, T. A. 2017. Environmental growth of *Enterococci* and *Escherichia coli* in feedlot catch basins and a constructed wetland in the absence of fecal input. *Environmental Science and Technology*, 51(10): 5386–5395. doi: 10.1021/acs.est.6b06274

USDA/FSIS. 2021. Safe and suitable ingredients used in the production of meat and poultry products. FSIS Directive 7120.1, Revision 56, USDA-FSIS. www.fsis.usda.gov/sites/default/files/media_file/2021-09/7120.1.pdf

USDA/FSIS. 2014. Pre-Harvest Management Controls and Intervention Options for Reducing Shiga Toxin-Producing Escherichia coli Shedding in Cattle: An Overview of Current Research [online]. USDA Guideline ID FSIS-GD-2014-0012. Cited 22 March 2022. www.fsis.usda.gov/guidelines/2014-0012?msclkid=5ff5ce26aab311ec b8cae5d20709d709

Van Donkersgoed, J., Jericho, K. W. F., Grogan, H. & Thorlakson, B. 1997. Preslaughter hide status of cattle and the microbiology of carcasses. *Journal of Food Protection*, 60: 1502–1508. doi: 10.4315/0362-028x-60.12.1502

Velasco, R., Ordóñez, J. A., Cambero, M. I. & Cabeza, M. C. 2015. Use of E-beam radiation to eliminate *Listeria monocytogenes* from surface mould cheese. *International Microbiology*, 18: 33–40. doi: 10.2436/20.1501.01.232

Venegas-Vargas, C., Henderson, S., Khare, A., Mosci, R. E., Lehnert, J. D., Singh, P., Ouellette, L. M., Norby, B., Funk, J. A., Rust, S., Bartlett, P. C., Grooms, D. & Manning, S. D. 2016. Factors associated with Shiga toxin-producing *Escherichia coli* shedding by dairy and beef cattle. *Applied and Environmental Microbiology*, 82: 5049–56. doi: 10.1128/AEM.00829-16

Venkitanarayanan, K. S., Zhao, T. & Doyle, M. P. 1999. Antibacterial effect of lactoferricin B on *Escherichia coli* O157:H7 in ground beef. *Journal of Food Protection*, 62: 747–750. doi: 10.4315/0362-028x-62.7.747

Verbeke, J., Piepers, S., Supré, K. & De Vliegher, S. 2014. Pathogen-specific incidence rate of clinical mastitis in Flemish dairy herds, severity, and association with herd hygiene. *Journal of Dairy Science*, 97: 6926–6934. doi: 10.3168/jds.2014-8173

Viazis, S., Farkas, B. E. & Jaykus, L. A. 2008. Inactivation of bacterial pathogens in human milk by high-pressure processing. *Journal of Food Protection*, 71: 109–118. doi: 10.4315/0362-028x-71.1.109

Vidovic, S. & Korber, D. R. 2006. Prevalence of *Escherichia coli* O157 in Saskatchewan cattle: Characterization of isolates by using random amplified polymorphic DNA PCR, antibiotic resistance profiles, and pathogenicity determinants. *Applied and Environmental Microbiology*, 72: 4347–4355. doi: 10.1128/aem.02791-05

Vosough Ahmadi, B., Frankena, K., Turner, J., Velthuis, A. G. J., Hogeveen, H. & Huirne, R. B. M. 2007. Effectiveness of simulated interventions in reducing the estimated prevalence of *E. coli* O157:H7 in lactating cows in dairy herds. *Veterinary Research*, 38(5): 755–771. doi: 10.1051/vetres:2007029

Walia, K., Argüello, H., Lynch, H., Grant, J., Leonard, F. C., Lawlor, P. G., Gardiner, G. E. & Duffy, G. 2017. The efficacy of different cleaning and disinfection procedures to reduce *Salmonella* and *Enterobacteriaceae* in the lairage environment of a pig abattoir. *International Journal of Food Microbiology*, 246: 64–71. doi: 10.1016/j. ijfoodmicro.2017.02.002

Wang, O., Liang, G., McAllister, T. A., Plastow, G., Stanford, K., Selinger, B. & Guan, L. L. 2016. Comparative transcriptomic analysis of rectal tissue from beef steers revealed reduced host immunity in *Escherichia coli* O157:H7 super-shedders. *PLOS ONE*, 11: e0151284. doi: 10.1371/journal.pone.0151284

Ward, L. R., Kerth, C. R. & Pillai, D. 2017. Nutrient profiles and volatile odorous compounds of raw milk after exposure to electron beam pasteurizing doses. *Journal of Food Science*, 82: 1614–1621. doi: 10.1111/1750-3841.13763

Wells, J. E., Shackelford, S. D., Berry, E. D., Kalchayanand, N., Bosilevac, J. M. & Wheeler, T. L. 2011. Impact of reducing the level of wet distillers grains fed to cattle prior to harvest on prevalence and levels of *Escherichia coli* O157:H7 in feces and on hides. *Journal of Food Protection*, 74: 1611–1617. doi: 10.4315/0362-028X. JFP-11-160

Wells, J. E., Berry, E. D., Kim, M., Shackelford, S. D. & Hales, K. E. 2017. Evaluation of commercial-agonists, dietary protein, and shade on fecal shedding of *Escherichia coli* O157:H7 from feedlot cattle. *Foodborne Pathogens and Disease*, 14: 649–655. doi: 10.1089/fpd.2017.2313

Wetzel, A. N. & LeJeune, J. T. 2006. Clonal dissemination of *Escherichia coli* O157:H7 subtypes among dairy farms in Northeast Ohio. *Applied and Environmental Microbiology*, 72: 2621–2626. doi: 10.1128/AEM.72.4.2621-2626

Wheeler, T. L., Kalchayanand, N., & Bosilevac, J. M. 2014. Pre- and post-harvest interventions to reduce pathogen contamination in the U.S. beef industry. *Meat science*, 98(3): 372–382. doi: 10.1016/j.meatsci.2014.06.026

Whittam, T. S., McGraw, E. A. & Reid, S. D. 1998. 5 Pathogenic *Escherichia coli* O157:H7: A model for emerging infectious diseases. *Biomedical Research Reports*, 1: 163–183. doi: 10.1016/S1874-5326(07)80029-9

Wisener, L. V., Sargeant, J. M., O'Connor, A. M., Faires, M. C. & Glass-Kaastra, S. K. 2015. The use of direct-fed microbials to reduce shedding of *Escherichia coli* O157 in beef cattle: A Systematic Review and Meta-analysis. *Zoonoses and Public Health*, 62: 75–89. doi: 10.1111/zph.12112

Wolf, M. J., Miller, M. F., Parks, A. R., Loneragan, G. H., Garmyn, A. J., Thompson, L. D. & Brashears, M. M. 2012. Validation comparing the effectiveness of a lactic acid dip with a lactic acid spray for reducing *Escherichia coli* O157:H7, *Salmonella*, and non-O157 Shiga toxigenic *Escherichia coli* on beef trim and ground beef. *Journal of Food Protection*, 75: 1968–1973. doi: 10.4315/0362-028x.jfp-12-038

Xu, Y., Dugat-Bony, E., Zaheer, R., Selinger, L., Barbieri, R., Munns, K., McAllister, T. A. & Selinger, L. B. 2014. *Escherichia coli* O157:H7 super-shedder and non-shedder feedlot steers harbour distinct fecal bacterial communities. *PLOS ONE*, 9: e98115. doi: 10.1371/journal.pone.0098115

Yalçin, S., Nizamlioglu, M. & Gürbüz, Ü. 2004. Microbiological conditions of sheep carcasses during the slaughtering process. *Journal of Food Safety*, 24(2): 87–93. doi: 10.1111/j.1745-4565.2004.tb00377.x

Yildirim, S., Rocker, B., Pettersen, M., Nilsen-Nygaard, J., Ayhan, Z., Rutkaite, R., Radusin, T., Suminska, P., Marcos, B. & Coma, V. 2018. Active packaging applications for food. *Comprehensive Reviews in Food Science and Food Safety*, 17: 165–199. doi: 10.1111/1541-4337.12322

Zaheer, R., Dugat-Bony, E., Holman, D., Cousteix, E., Xu, Y., Munns, K., Selinger, L. J., Barbieri, R., Alexander, T., McAllister, T. A. & Selinger, L. B. 2017. Changes in bacterial community composition of *Escherichia coli* O157:H7 super-shedder cattle occur in the lower intestine. *PLOS ONE*, 12: e0170050. doi: 10.1371/journal.pone.0170050

Zhao, T., Tkalcic, S., Doyle, M. P., Harmon B. G., Brown, C. A. & Zhao, P. J. 2003. Pathogenicity of enterohemorrhagic *Escherichia coli* in neonatal calves and evaluation of fecal shedding by treatment with probiotic *Escherichia coli*. *Journal of Food Protection*, 66: 924–30. doi: 10.4315/0362-028x-66.6

Zhilyaev, S., Cadavez, V., Gonzales-Barron, U., Phetxumphou, K. & Gallagher, D. 2017. Meta-analysis on the effect of interventions used in cattle processing plants to reduce *Escherichia coli* contamination. *Food Research International,* 93: 16–25. doi: 10.1016/j.foodres.2017.01.005

Zhou, M., Hünerberg, M., Chen, Y., Reuter, T., McAllister, T. A., Evans, F., Critchley, A. T. & Guan, L. L. 2018. Air-dried brown seaweed, *Ascophyllum nodosum*, alters the rumen microbiome in a manner that changes rumen fermentation profiles and lowers the prevalence of foodborne pathogens. *mSphere,* 3(1):e00017-18. doi: 10.1128/mSphere.00017-18

Zhou, Y., Karwe, M. V. & Matthews, K. R. 2016. Differences in inactivation of *Escherichia coli* O157:H7 strains in ground beef following repeated high-pressure processing treatments and cold storage. *Food Microbiology,* 58: 7–12. doi: 10.1016/j.fm.2016.02.010

FAO/WHO Microbiological Risk Assessment Series

19 *Salmonella* and *Campylobacter* in chicken meat: meeting report, 2009

20 Risk assessment tools for *Vibrio parahaemolyticus* and *Vibrio vulnificus* associated with seafood: meeting report, 2020

21 *Salmonella* spp. in bivalve molluscs: risk assessment and meeting report, in press

22 Selection and application of methods for the detection and enumeration of human pathogenic halophilic *Vibrio* spp. in seafood: guidance, 2016

23 Multicriteria-based ranking for risk management of foodborne parasites, 2014

24 Statistical aspects of microbiological criteria related to foods: a risk managers guide, 2016

25 Risk-based examples and approach for control of *Trichinella* spp. and *Taenia saginata* in meat: meeting report, 2020

26 Ranking of low-moisture foods in support of microbiological risk management: meeting report and systematic review, 2022

27 Microbiological hazards in spices and dried aromatic herbs: meeting report, 2022

28 Microbial safety of lipid based ready-to-use foods for management of moderate acute malnutrition and severe acute malnutrition: first meeting report, 2016

29 Microbial safety of lipid based ready-to-use foods for management of moderate acute malnutrition and severe acute malnutrition: second meeting report, 2021

30 Interventions for the control of non-typhoidal *Salmonella* spp. in beef and pork: meeting report and systematic review, 2016

31 Shiga toxin-producing *Escherichia coli* (STEC) and food: attribution, characterization, and monitoring; report, 2018

32 Attributing illness caused by Shiga toxin-producing *Escherichia coli* (STEC) to specific foods: report, 2019

33 Safety and quality of water used in food production and processing: meeting report, 2019

34 Foodborne antimicrobial resistance: role of the environment, crops and biocides: meeting report, 2019.

35 Advance in science and risk assessment tools for *Vibrio parahaemolyticus* and *V. vulnificus* associated with seafood: meeting report, 2021.

36 Microbiological risk assessment guidance for food: guidance, 2021

37 Safety and quality of water used with fresh fruits and vegetables, 2021

38 *Listeria monocytogenes* in ready-to-eat (RTE) foods: attribution, characterization and monitoring: 2022

39 Control measures for Shiga toxin-producing *Escherichia coli* (STEC) associated with meat and dairy products: meeting report, 2022